Introduction
to
Biological Physics

Introduction
to
Biological Physics

M. E. J. Holwill and N. R. Silvester

Physics Department
Queen Elizabeth College
University of London

JOHN WILEY & SONS
London • New York • Sydney • Toronto

Library of Congress catalog card number
72–8604

ISBN 0 471 40863 8   Cloth bound
ISBN 0 471 40864 6   Paper bound

Photosetting by Thomson Press (India) Limited.

Printed in Great Britain by Unwin Brothers Ltd., The Gresham Press, Old Woking, Surrey.

Preface

Some years ago we began to teach a forty–lecture course of 'Physics for Biology' to a mixed group of first–year undergraduate students. Their major fields of study were biology, microbiology, physiology or nutrition; physics was a strictly ancillary activity which had to be shown to be 'relevant'. What we tried to give the students was some familiarity with, and understanding of, those branches of physics which they might need in a modern biological laboratory or encounter in the literature of quantitative biology and biophysics.

With such an aim, it was natural that much of our lecture material could be extracted only from advanced (and expensive) treatises on single topics, such as electron microscopy, and our students deplored the absence of a single textbook that would cover the course. We discovered that most ancillary physics texts aim to teach a diluted version of the whole corpus of undergraduate physics, while we were attempting a somewhat different exercise—to produce informed biologists rather than 'mini–physicists'.

This book is therefore written to cover some of that unclaimed territory whose borders merge with those of elementary physics, biological instrumentation and biophysics. If our readers become familiar with the landmarks of this territory and learn enough of the language to be able to 'ask their way', the book will have achieved its purpose.

We are grateful to those who have given us permission to use diagrams and photographs, and we acknowledge them individually in the text.

We are also grateful to our wives for much patient forbearance during the writing of this book, and especially to Gillian Silvester, who deciphered and then typed from the manuscript.

NORMAN SILVESTER
MICHAEL HOLWILL

August 1972

Contents

Introduction

The student of contemporary biology finds it increasingly necessary to adopt a quantitative approach to the subject. This applies not only to experimental aspects (such as the use of electronic instrumentation) but also to the basic theory in such topics as, for instance, photosynthesis and molecular genetics. In many aspects, the recent development of quantitative biology has depended on the fundamental principles of the physical sciences, and some grasp of these principles is essential in order to appreciate fully the concepts of modern biological science.

In the following chapters we shall develop those physical concepts and their applications which, in our opinion, are likely to be of most relevance and benefit to you, either immediately or during your future work. The topics included are only treated to the depth which will enable you to follow articles in biological journals or continue your study of the subject in more specialized texts. Of necessity, the number of topics is only a selection from the many that are possible, and we have had to omit those which should be encountered in physical chemistry or mathematics courses.

We assume as our starting point that you have the elementary knowledge and understanding of physics and mathematics that would be gained in 'O' level G.C.E. courses, although in some chapters the elementary principles are briefly revised at the beginning. We use the International System of Units (S.I.) throughout this book, except where (as with the Ångstrom unit in electron microscopy) current practice and the literature show little signs of adopting the modern convention. The system of S.I. units is similar in most respects to the M.K.S. (metre–kilogram–second) system which has been in extensive use for many years.

Specific references for further reading are included at the end of each chapter, but there are several books which are relevant to biological physics as a whole, and others which will introduce the reader to topics we have been unable to include. We list a selection of these books below.

We hope that when you have read the following chapters and worked through the problems, you will be better able to appreciate the application of physical principles in biological science.

General Reading

Ackermann, E. (1962). *Biophysical Science*. Prentice–Hall, New York.

Barry, J.M., and E.M. Barry (1969). *An Introduction to the Structure of Biological Macromolecules*. Prentice–Hall, New Jersey.

Bergner, J., E. Gelbke and W. Mehliss (1966). *Practical Photomicrography*. Focal Press, London.

Chapman, D., and R.B. Leslie (1967). *Molecular Biophysics*. Oliver and Boyd, Edinburgh.

Dickerson, R.E., and I. Geis (1969). *The Structure and Action of Proteins*. Harper and Row, New York.

Engel, C.E. (Ed.) (1968). *Photography for the Scientist*. Academic Press, New York.

Finean, J.B. (1967). *Engstrom–Finean Biological Ultrastructure*, 2nd ed. Academic Press, New York.

Haggis, G.H., and others (1964). *Introduction to Molecular Biology*. Longmans, London.

Haynes, R.H., and C. Hanawalt (1968). *The Molecular Basis of Life* (readings from *Scientific American*). Freeman, San Francisco.

Hughes, M.N., A.M. James and N.R. Silvester (1970). *S.I. Units and Conversion Tables*. Machinery Publishing Co., Brighton.

Jarman, M. (1970). *Examples in Quantitative Zoology*. Arnold, London.

Kay, R.H. (1964). *Experimental Biology*. Chapman and Hall, London.

Marriott, F.H.C. (1970). *Basic Mathematics for the Biological and Social Sciences*. Pergamon, Oxford.

Richards, J.A., F.W. Sears, M. Wehr and M.W. Zemansky (1960). *Modern University Physics*. Addison–Wesley, Reading, Mass.

Royal Society (1971). *Quantities, Units and Symbols*. The Royal Society, London.

Setlow, R.B., and E.C. Pollard (1962). *Molecular Biophysics*. Addison–Wesley, Reading, Mass.

Smith, A.W., and J.N. Cooper (1964). *Elements of Physics*, 7th ed. McGraw–Hill, New York.

CHAPTER 1
Mechanics and the
Properties of Solids

Forces play a vital role in the activity of any biological system. For example, cohesive forces bind the various parts of an organism together, electrostatic forces are dominant in the chemical reactions which occur continuously in cells, while frictional, hydrodynamic and other forces are responsible for movement. Living systems also use up energy. Chemical energy is used just to keep the cells of an organism alive; more energy is, of course, necessary if the organism moves or participates in other activities.

The concepts of force and energy are basic to an understanding of the behaviour of molecules, cells and organisms. In this chapter, therefore, we shall try to give the reader an understanding of these concepts by illustrating their application to solids and to systems which are of biological interest.

1.1 Forces

Suppose we have a block of wood resting on a highly polished table. The block will remain at rest unless we push or pull it, lift it up or disturb it in some other way. To move the block it is necessary to exert a *force* upon it. Suppose that we give the block a sharp tap in such a way that it glides across the table. If the table is long and sufficiently highly polished, the block will slide in a straight line for a comparatively long time with little reduction in speed. If we want to alter its speed (e.g. slow it down) or change the direction of its path, it is again necessary to exert a *force* upon the block. In summary, we can say that a body will remain at rest, or in motion at constant speed in a straight line, unless it is compelled to change its state by the action of a *force*.

The magnitude of a force is estimated from the effect it has upon the body on which it acts. Experimentally it is found that if a mass m accelerates at a rate a the force F which produces this acceleration is given by $F \propto m \times a$. The S.I. unit of force is the newton (symbol N), which is defined as the net force necessary to cause a body having a mass of one kilogramme to accelerate at a rate of one metre per second per second (1 m s^{-2}) along the direction in which the force is applied. If all quantities are expressed in S.I. units, the force equation then becomes

$$F = m \times a. \tag{1.1}$$

1

The newton is a 'derived' unit which, as one sees from the right–hand side of equation (1.1), can be expressed by the product kg m s^{-2}, in terms of the fundamental S.I. units.

1.1.1 *Moment of a Force*

Many situations exist where a force is used to rotate a body about a pivot. In the reptilian jaw (Figure 1.1a), for example, the jaw muscle exerts a

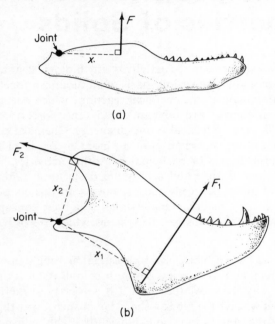

Figure 1.1.   Probable lines of action of forces which acted to close the lower jaws of certain extinct reptiles. F, F_1 and F_2 represent the forces used while x, x_1 and x_2 are the perpendicular distances from the forces to the pivot or joint about which they rotate the jaw. (Adapted with kind permission from A.W. Crompton, *Proc. Zool. Soc. Lond.*, **140**, 699–753, 1963)

force which rotates the lower jaw about the joint, thereby closing the mouth of the reptile. The magnitude of the turning effect is indicated by what is called the *moment* of the force, which is the product of the force and the perpendicular distance from the pivot to the line of action of the force. In Figure 1.1a the moment of the force is $F \times x$. The jaw of the reptile is used to crush its food, so that the moment of the force F can be regarded as a measure of how effective the jaw will be for this process. In Figure 1.1b two muscles are used to close the jaw. The overall moment in this case is simply the sum of the moments of the two separate forces, i.e. $F_1x_1 + F_2x_2$. Clearly, the larger the values for F and x the more effective will be the crushing action of the jaw. For a given muscle to achieve the greatest effectiveness in the crushing activity, its line of action should be as far away from the joint as possible.

1.1.2 *Couples*

In some cases a body is acted upon by two equal parallel forces, oppositely directed but with different lines of action, as shown in Figure 1.2. This

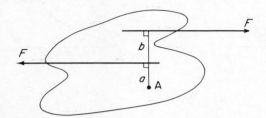

Figure 1.2. Two equal parallel forces acting on a body constitute a couple; a and $a+b$ are the perpendicular distances from an arbitrary point A to the lines of action of the two forces

system of forces constitutes a couple (or torque). The sum of the moments of the two forces about any arbitrary point such as A is

$$F(a+b) - Fa = Fb. \tag{1.2}$$

Notice here that we have reckoned the moment as positive if it would tend to produce a clockwise rotation about the point considered, while the moment of a force which would rotate the body in an anticlockwise direction is reckoned as negative. The quantity $c = F \times b$ is independent of the position of the point A and is known as the *moment of the couple*.

If this couple were continuously applied to the body it would produce an accelerating rate of rotation, analogous to the acceleration in speed produced by a direct force. An equation similar to (1.1) applies in this case:

$$c = J a_\theta, \tag{1.3}$$

where a_θ is the angular acceleration in radians per second per second (rad s^{-2}) and J is a quantity called the moment of inertia of the body (see Section 1.3.2).

1.1.3 *Vector Representation of Forces*

A force is an example of a *vector* quantity, that is one which requires a knowledge of a direction as well as a magnitude for its complete specification. Other examples of vector quantities are velocity and acceleration. In contrast, a *scalar* quantity, such as temperature or length, is one which is completely described by a single number indicating its magnitude. A vector quantity may be conveniently represented by a straight line drawn parallel to the line of action of the quantity (e.g. along the direction in which a force is acting). The length of the line is made proportional to the magnitude of the quantity and the direction in which the quantity is acting is shown by an arrow at the head of the line. Thus, in Figure 1.3a the length of the line OP

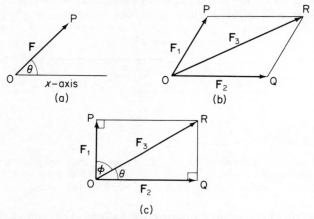

Figure 1.3. (a) The length and direction of OP represent the magnitude and direction of the vector **F**. (b) **F**₃ is the *resultant* vector of two vectors **F**₁ and **F**₂. (c) **F**₁ and **F**₂ are *components* in the appropriate directions of the vector **F**₃

represents the magnitude of a force, **F**, acting at a point O in a direction at an angle θ to the x–axis. Notice that the symbol **F** which we have just used for force is printed in bold type. This is a conventional way of indicating that we are considering a vector quantity and we shall use this notation from now on. The *magnitude* of a vector quantity is represented by the same letter but in italic type (F) to show that we are not concerned with the direction.

Usually the effect of two forces acting on a body can, unless the forces create a couple, be reproduced by a single force. The magnitude and direction of the single force which is equivalent to the two forces can easily be found by the parallelogram law for the addition of vectors. In Figure 1.3b, OP and OQ represent two vectors **F**₁ and **F**₂. The single equivalent vector (or *resultant* as it is called), **F**₃, is found by completing the parallelogram OPRQ of which OP and OQ form adjacent sides. The resultant is the vector OR. Conversely, it is sometimes useful to represent a single vector by two separate vectors. This technique, known as resolving the vector, is shown in Figure 1.3c, which represents a similar situation to that of

Figure 1.3b except that the vectors F_1 and F_2 are equivalent in effect to the single vector F_3. The vector F_2 alone will produce the same effect in the direction OQ as would the vector F_3, since F_1, acting at right angles to OQ, gives no additional effect along that direction. We say that F_2 is the *component* of F_3 in the direction of OQ. Similarly, F_1 is the component of F_3 in the direction of OP. The magnitudes of F_1 and F_2 can be seen to be $F_3 \cos \phi$ and $F_3 \cos \theta$ respectively. Thus, to obtain the total effect or component of a vector in a particular direction, we draw a line from the arrowed end of the vector perpendicular to a line drawn in the required direction which passes through the other end of the vector. In the triangle so formed (e.g. ORQ in Figure 1.3c) the required vector length (OQ) is called the projection of the original vector length (OR), and represents the magnitude of the component vector.

1.1.4 *Impulse and Momentum*

If a net force acts for a length of time on a body that is initially at rest, we can calculate the velocity V of the body at any instant, since we know the acceleration of the body from equation (1.1). The increase in velocity during a short time dt is $a\ dt$, assuming the acceleration a changes very little over this time. From equation (1.1) we have

$$F\ dt = m\,a\ dt$$
$$= m\ dV$$

where dV is the increase in velocity due to F acting over the time dt. Over the total time for which F is acting it may vary considerably, but we can assume the value of F to be constant if we replace it by the average value, $\bar{F}$. Taking together all the intervals of time dt that make up the total time t in which we are interested, and adding together all the corresponding increases in velocity, we obtain

$$\bar{F}t = mV. \tag{1.4}$$

From this equation we can find the velocity of the body at any time if we know the average force that has acted on it. The left–hand side of the equation is the product of a given force acting over a given time, and is called the *impulse* imparted to the body. From equation (1.4) we see that the product of the mass and velocity, called *momentum*, is always equal to the total impulse the body has received. The greater the mass, the smaller is the velocity that results from a given impulse; this is shown by, for example, the different effects of hitting a golf ball and a table–tennis ball with a table–tennis bat.

The momentum of a body is a vector quantity, that is it has a direction, since the product mV includes the velocity of the body. It can be shown to follow from the laws of motion that in interactions (such as collisions) between several bodies, the total momentum of the bodies is the same before and after the interaction (i.e. momentum is conserved). To take a simple example, suppose an ice–hockey player of mass 90 kg moving at 5 metres per second is tackled by another player of mass 80 kg moving at right angles to his path with a speed of 7·5 metres per second. The two players, locked together, slide on the ice—in what direction and with what speed? The momentum of the first player is $90 \times 5 = 450$ kg m s^{-1}; that of the second is $80 \times 7·5 = 600$ kg m s^{-1} at right angles to the first value. Combining the two values as vectors, we find the total momentum before the collision is 750 kg m s^{-1} at an angle with tangent 600/450 to the path

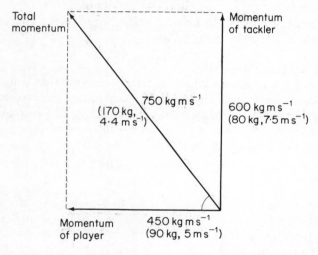

Figure 1.4.   Vector addition of the momenta of two ice–hockey players

of the first player (Figure 1.4). Since the combined mass of the players is 170 kg, their velocity after the tackle is $750/170 = 4·4$ m s^{-1} in the direction of the total momentum.

1.1.5 Forces in Equilibrium

A body which is at rest or which is moving at a constant speed in a straight line is said to be in equilibrium. The *net* force acting on such a body is zero. In practice a number of forces will be acting on the body. First there will be its weight, which is a force acting vertically downwards produced by the gravitational attraction between the body and the earth. In the simple case of a wooden block resting on a table (Figure 1.5) there is a force of reaction, **R**, acting vertically upwards on the block. Such a force must exist, otherwise nothing would prevent the body from passing downwards through the table. It can be shown that the magnitude of the reaction is equal to the weight of the body and that the line of action of the reaction is the same as that of the weight. If the lines of action were different (e.g. **R′** in Figure 1.5) the forces would rotate the body, which would therefore not be in equilibrium.

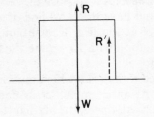

Figure 1.5.   A body resting on a surface is a simple example of a body in equilibrium. The reaction **R** is equal to the weight (**W**) of the body. The force **R′** is discussed in the text

Again, if the lines of action were the same but the forces were different in magnitude, the body would move either upwards or downwards, depending on whether **R** was greater or less than **W**.

There is a set of conditions which can be applied to the known forces acting on a body to enable us to determine whether or not the body should be in equilibrium. The first condition is that the sum of the components of the forces, resolved in any two chosen directions, must be zero if the body is to be in equilibrium. This is simply the condition that no net force should act on the body to produce acceleration. The directions in which the forces are resolved need not be at right angles, although it is frequently convenient to make them so.

The second condition satisfied by a system of forces in equilibrium is that the sum of the moments of all the forces about any point is zero. This is simply the condition that no rotation of the body can occur if it is in equilibrium.

There exist alternative expressions of the conditions we have mentioned. For example, a body will be in equilibrium if the sum of the moments of all

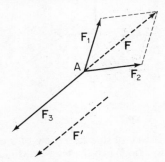

Figure 1.6.   If a body is in equilibrium under the action of three forces (**F**$_1$, **F**$_2$, **F**$_3$), the lines of action of the forces must all pass through a single point. **F** is the resultant of forces **F**$_1$ and **F**$_2$. The force **F′** is discussed in the text

the forces about any three points which are not in a straight line is zero.

It is fairly common to find a body in equilibrium under the action of three forces (Figure 1.6). In this case, the lines of action of all three forces must pass through a common point (A in Figure 1.6). This is so since the resultant of, say, forces $\mathbf{F}_1$ and $\mathbf{F}_2$ must be equal to and have the same line of action as the force $\mathbf{F}_3$. If $\mathbf{F}_3$ had a different line of action (e.g. $\mathbf{F}'$ in Figure 1.6), the resultant of $\mathbf{F}_1$ and $\mathbf{F}_2$ would, in combination with $\mathbf{F}_3$, rotate the body, so that it would not be in equilibrium.

The conditions for equilibrium can be used to indicate which muscles are used for particular operations by animals. As an example, let us consider the action of the lower jaw in a carnivore such as *Martes*, the marten (Figure 1.7). These animals use their carnassial teeth, which are placed well back along the jaw bone, to cut through flesh. A force $\mathbf{P}$, roughly perpendicular to the jaw, is exerted on the lower jaw in the cutting action. The force $\mathbf{P}$ is

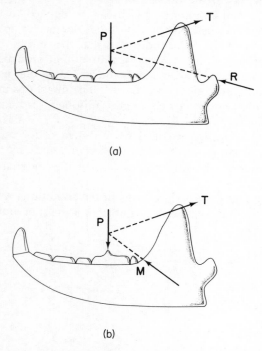

(a)

(b)

Figure 1.7. Forces acting on the lower jaw of a carnivore, *Martes*, during a cutting action. (a) When the temporalis muscle only is used, equilibrium is maintained by a reaction $\mathbf{R}$ at the jaw articulation. (b) When both temporalis and masseter muscles are used, little or no reaction at the jaw articulation is needed to maintain equilibrium. (Adapted with kind permission from J.M. Smith and R.J.G. Savage, *J. Linn. Soc. (Zool.)*, **42**, 603–622, 1956)

one of reaction, which arises because the flesh is being compressed between the upper and lower jaws. Two muscles, the temporalis and masseter, can be used to produce the force opposing **P**, thereby maintaining equilibrium. If the temporalis muscle were used alone, as shown in Figure 1.7a, equilibrium would be maintained by means of a third force, the reaction **R** at the jaw articulation. This force is of such a magnitude and in such a direction that it might dislocate the jaw. If, on the other hand, both the masseter and the temporalis muscles are used (Figure 1.7b), the equilibrium is possible with little or no reaction at the articulation, and this appears to be the situation in practice.

Analyses of this type have been used to explain the fact that in carnivores the temporalis muscle is, relatively, much heavier than in the herbivores. On the other hand, the pterygoideus muscle (one which produces transverse 'grinding' movements of the lower jaw) is (again relatively) much heavier in herbivores than the corresponding muscle in carnivores.

1.1.6 *Machines*

There are, of course, many different forms of machine, but all are in essence used to convert one force (F_i) into another force (F_o); F_o is more suitable in magnitude or in line of action (or in both) than F_i to perform the task for which the machine is designed. The ratio F_o/F_i is called the *mechanical advantage* of the machine. Many parts of vertebrates, particularly the limbs, are examples of a simple machine, the lever. Figure 1.8a, for instance, shows the bones of the human arm together with the muscles

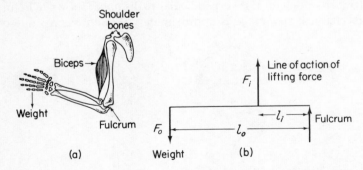

Figure 1.8. (a) Diagram of the human arm and muscle used to lift a weight. (b) Idealized representation of (a); Fulcrum is at head of arrow

which are used when a weight is lifted. The fore–arm is acting as a lever which may be represented more simply by Figure 1.8b. The elbow joint is the *fulcrum* about which the lever moves, while F_i is the 'vertical' component of the force exerted by the muscle (the horizontal component is balanced by a reaction at the elbow). If the system is in equilibrium and there is no friction at the fulcrum, then by taking moments about the fulcrum we have

$$\mathbf{F}_i\, l_i = \mathbf{F}_o\, l_o \tag{1.5}$$

or

$$\frac{\mathbf{F}_o}{\mathbf{F}_i} = \frac{l_i}{l_o}. \tag{1.6}$$

Clearly the force applied to the arm by the muscle must be larger than the weight to be lifted. Equation (1.6) shows that, in principle, it is possible to calculate the mechanical advantage of the system by making measurements on the arm itself. It is, in fact, a feature of any machine that a theoretical mechanical advantage can be calculated from its dimensions. (In many machines a movement of the applied force, $\mathbf{F}_i$, produces a movement of the output force, $\mathbf{F}_o$. The ratio of the distance moved by the applied force to that moved by the output force is called the *velocity ratio* and is easily shown to be equal to the theoretical mechanical advantage.) The practical mechanical advantage is never as great as the theoretical one because the applied force has to overcome frictional forces in addition to the load itself. In the example of the human arm, the frictional forces arise in the elbow joint.

The value of the mechanical advantage depends on the machine's function. To illustrate this point, let us consider the example of the action of the fore–limbs of two very different animals, the armadillo and the horse. The fore–legs of the armadillo are used for digging, an operation which requires a reasonably large force. The muscular force mainly used for digging acts upwards along the line perpendicular to distance l shown in Figure 1.9(a). The mechanical advantage is approximately l/h and has a value of about

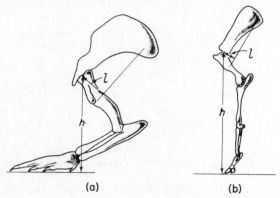

(a)   (b)

Figure 1.9.   Skeletons of the fore–legs of (a) an armadillo and (b) a horse to show the difference in the mechanical advantage of the muscle used for digging. (Adapted with kind permission from J.M. Smith and R.J.G. Savage, *School Science Review*, **40**, 289–301, 1959)

1/4 for the armadillo. A horse uses the corresponding muscle when it is running. The theoretical mechanical advantage, and hence the velocity ratio (l/h in Figure 1.9b), is about 1/13, a value significantly smaller than that for the armadillo. (The actual values obtained in each case will depend somewhat on the position of the limb.) The low velocity ratio means that a relatively large movement of the horse's foot results when the muscle contracts, and a fast running speed is hence possible. The mechanical advantage, although low, is sufficient to allow the horse to propel itself by this muscular action. The higher mechanical advantage, and hence the higher velocity ratio, characteristic of the armadillo's fore–limbs means that rapid movement is not easily achieved. In addition to the mechanical advantage, other factors, such as the moment of inertia (Section 1.3.2) of the limb, are of importance when considering activities such as running and digging. It is generally found, however, that the mechanical advantage of the fore–legs of digging animals is greater than that of running animals. Similar adaptations of mechanical advantage to a particular task are also found in other animals and in other muscles.

1.2 Work and Energy

In the language of physics, we say that work is done (i.e. energy is expended) when a force moves its point of application. If, as in Figure 1.10, the point

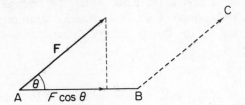

Figure 1.10. Illustrating the parameters involved in the discussion of a moving force. AB is the direction of movement of the point of application of the force, while BC is its final line of action

of application of a force **F** moves through a distance x which is not along the line of the force, the energy expended, W, is given by

$$W = F x \cos \theta. \tag{1.7}$$

Notice that the force is resolved along the direction of movement to obtain the component ($F \cos \theta$) acting in the direction of x. The S.I. unit of work is the joule (J) and is the energy expended when a force of one newton moves through one metre along its own line of action. The rate of energy expenditure is called power and can be measured in joules per second. However, one joule per second, the unit of power, is given a distinctive name, the watt (W). (Thus $1\text{ W} = 1\text{ J s}^{-1}$.)

Two types of energy are recognized: potential energy and kinetic energy. Potential energy may be understood by considering a mass of m kilogrammes held at a height of h metres above a surface. We can say that the mass has potential energy because if it were released it would be capable of transferring energy to another body. It could, for example, raise a second mass attached to it by a rope which ran through a pulley.

To calculate the value of the potential energy it is necessary to know the force which acts on the body during its fall. If the body were allowed to fall freely it would accelerate at a rate $\mathbf{g}$ (in m s^{-2}), where $\mathbf{g}$ is the acceleration due to gravity. The force in newtons acting vertically downwards on the body is then, by equation (1.1), $m\,\mathbf{g}$. The force which produces the acceleration $\mathbf{g}$ is a gravitational attraction between the earth and the body and remains even if the body is subjected to other restraints. The force, $m\,\mathbf{g}$, which acts vertically downwards is the *weight* of the body. Weight is thus a vector quantity whereas mass, which represents the amount of matter in a body, is a scalar quantity.

The amount of work which would be done by the mass in falling h metres is $m\,g\,h$ joules, and we therefore say that the potential energy of the mass when at the height of h metres is $m\,g\,h$ joules. The units of energy are therefore the same as those of work.

Kinetic energy, as the name implies, is the energy which a body possesses by virtue of its motion. It is the work which the body could do on being brought to rest. If the mass of the body is m kilogrammes and its velocity is $\mathbf{V}$ metres per second, the kinetic energy E_k can be shown to be given by

$$E_k = \tfrac{1}{2}\,m\,V^2. \tag{1.8}$$

When a system does work, it uses up energy. In accordance with the principle of conservation of energy, the amount of energy which is used is equal to the work done by the system. Not all of the energy used, however, is converted into useful work. When the elbow is bent to lift a weight, for example, muscular (chemical) energy is used. Some of the energy is used to lift the weight, but some is used in overcoming frictional forces at the elbow joint and within the muscle itself. The energy lost in overcoming friction is transformed into heat energy. A useful measure of the performance of a machine is its efficiency, which is defined as the ratio of the amount of useful work done by the machine to the total energy expended by the machine in doing this work. It is possible, for instance, to measure the efficiency of muscular effort, which really represents the efficiency of conversion of chemical energy into mechanical work by a muscle. By using a bicycle ergometer, which measures the work done by the human leg muscles in pedalling, the maximum efficiency of these muscles has been found to be just over 20 per cent.

1.3 Movement in a Circle

Up to now we have limited our discussion mainly to the motion of bodies in straight lines. However, it is important to understand what happens when a body moves in a circle or circular arc, as such motions occur in that useful machine, the centrifuge, and in various muscular operations such as the beating of wings in flight.

We may describe the movement of the body by specifying the radius, r, of the orbit or circle in which it moves and the speed, V, of the body at every instant (Figure 1.11). As an alternative to the speed we may specify the

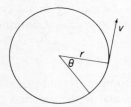

Figure 1.11.   Parameters involved in circular motion, where v is the speed of a point moving in a circular orbit of radius r and θ is the angle made by the radius with a reference line

angular velocity of the radius that links the body to the centre of the circular orbit. By angular velocity we mean the angle in radians which would be swept out by the radius in one second. If the body moves with a constant speed and orbits f times in one second then, since the radius sweeps out 2π radians on moving once around the circle, the angular frequency, ω, is given by

$$\omega = 2\pi f. \tag{1.9}$$

The speed of the body, again supposed to be constant, is found by dividing the circumference $(2\pi r)$ of the circle by the time taken to complete the orbit once. This time is just $2\pi/\omega$, so the speed is given by

$$V = r\omega. \tag{1.10}$$

Since the direction of movement of the body is continually changing, so also is its *velocity* (the vector quantity), although the *speed* (the magnitude of the vector) may remain constant. Since the velocity is changing a force must be acting on the body at all points along its path. To find the magnitude of this force consider Figure 1.12, which represents the body at two positions separated by a time dt and an angle $d\theta$. We need to determine the acceleration of the body. By resolving the velocity of the body in its second position along and perpendicular to the radius in the first position we see that the change in velocity is $V \sin(d\theta)$ towards the centre of the circle and $V(1 - \cos(d\theta))$ in a direction perpendicular to the radius. For small angles $\cos(d\theta)$ is unity, so that no change in velocity occurs perpendicular to the radius. Again, for small angles, $\sin(d\theta) = d\theta$, so the rate of change of velocity (i.e. the acceleration) is $V(d\theta/dt)$. Now $d\theta/dt$ is the angular velocity, ω,

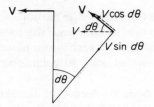

Figure 1.12. Showing two positions of a body moving in a circular orbit. This figure is discussed fully in the text

and hence, using equation (1.10), when a body moves in a circle it undergoes an acceleration $r\omega^2$ towards the centre of the circle. The inwardly directed (centripetal) force, $\mathbf{F}_c$, required to keep the body moving in its circular path is equal to the mass m of the body multiplied by its acceleration (equation 1.1) and has a magnitude given by

$$F_c = m\,r\,\omega^2. \tag{1.11}$$

The apparent force exerted by the body in attempting to continue its path in a straight line is called the centrifugal force, and is the reaction to the centripetal force exerted on it. The centrifugal force is thus also given by equation (1.11), but acts outwards.

1.3.1 *The Centrifuge*

If a container of liquid moves on the circumference of a circle, the molecules of the liquid and any particles suspended in it all experience the centrifugal forces described above. Thus as far as the container is concerned, any molecule within it appears to be acted on by a centrifugal force given by $m\,r\,\omega^2$ (equation 1.11), tending to move the molecule towards the side of

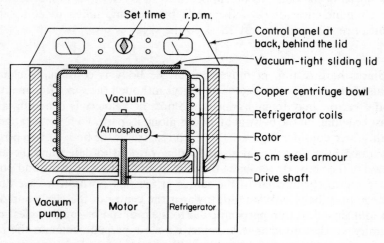

Figure 1.13. Diagram of a typical medium–sized preparative ultracentrifuge

the container furthest from the centre of the circle. In a stationary liquid the force acting to bring the particle to the bottom is m **g**, the force due to gravity. By placing the container in an appropriate machine, the centrifuge (Figure 1.13), we can create values of $r\omega^2$ much greater than g and artificially increase the force which causes suspended particles to sediment out of a liquid. For instance, if $r\omega^2 = 1000\ g$, the effective 'weight' of the particles in the centrifuge (acting of course outwards, not downwards) is a thousand times greater than their normal, gravitational weight.

In the typical bench centrifuge, a small electric motor spins a cradle holding ordinary test–tubes, and produces radial accelerations $(r\omega^2)$ of up to 5000 g. The *preparative ultracentrifuge* is used in biochemistry for such jobs as sedimenting parts of cells (e.g. mitochondria) out of suspension. In this machine, the liquid is contained in special plastic tubes placed in a solid metal casting (the rotor, Figure 1.14) which may be rotated at fre–

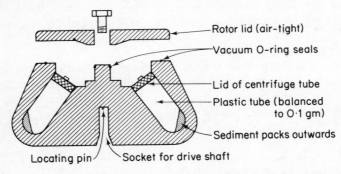

Figure 1.14. Diagram of a rotor used in a preparative ultracentrifuge

quencies up to 65,000 r.p.m. For reasons which will become apparent, the sealed rotor is spun in an evacuated centrifuge chamber which has an armoured steel casing.

It is instructive to consider the engineering problems posed in the operation of the ultracentrifuge. Let us take, for example, a rotor of diameter 30 cm which rotates at 30,000 r.p.m. and calculate the radial acceleration and the velocity of the rotor surface. The frequency of rotation is $30{,}000/60 = 500\ \mathrm{s}^{-1}$ and the angular frequency is 2π times this figure, i.e. $\omega = 1000\,\pi$ rad s^{-1}. The radial acceleration $r\omega^2$ is $(0{\cdot}30/2)\,10^6\,\pi^2$, or about $1{\cdot}5 \times 10^6\ \mathrm{m\ s}^{-2}$. Since g is $9{\cdot}81\ \mathrm{m\ s}^{-2}$, we find the acceleration of the rotor surface is about 150,000 g. To put this in practical terms, it means that each ounce of metal at the surface experiences an outward force of 150,000 oz, or about $4\frac{1}{2}$ tons weight. One problem, therefore, is to design centrifuge rotors which do not explode outwards when rotated at high speeds; the armour casing of the centrifuge protects the operator if a rotor failure occurs. The velocity of the rotor surface is given by $r\omega$ or $0{\cdot}15 \times 1000\,\pi$ m s^{-1}. This velocity is about 470 m s⁻¹ or 1700 km per hour, which is supersonic; to minimize the consequent air drag and surface heating, the rotor is spun in a vacuum chamber.

Because of the problems we have mentioned, the safe size of a centrifuge rotor is dictated by the maximum speed at which it is to run. For instance, a 60,000 r.p.m. rotor may only take a total liquid volume (distributed in several tubes) of 80 cm³ while a 20,000 r.p.m. rotor might accept 1000 cm³. To treat very large volumes of liquid (such as the culture media in which bacteria are grown) it is inconvenient to reload the rotor continually, and for this reason *continuous-flow* rotors have been designed. In such a device, the suspension enters the hollow, spinning, rotor continuously through a central channel, the sediment collects on the inner walls of the rotor and the clear liquid or supernatant is ejected through another channel by centrifugal pressure.

The use of the *analytical ultracentrifuge* for measurement of molecular masses, etc., is described in a reference at the end of the chapter.

1.3.2 *Moment of Inertia*

A body rotating about a fixed axis possesses kinetic energy, even though there is no net translational movement of the body. While for linear move-ment the kinetic energy is given by $E_k = \frac{1}{2}mV^2$, the kinetic energy for rotational movement has an analogous expression,

$$E_k = \tfrac{1}{2} J \omega^2, \tag{1.12}$$

where J is the *moment of inertia* of the body; the value of E_k here represents the amount of work which would be required to stop the rotational movement completely.

Let us calculate the kinetic energy of an insect wing (Figure 1.15) to see how the moment of inertia is related to the geometry of the wing. Consider

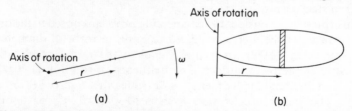

Figure 1.15.   Rotation of an insect wing at an angular velocity ω: (a) edge-on view, (b) top view

a very small strip of the wing at a distance r from the axis of rotation. If we consider the movement of the wing over a small time we can calculate its kinetic energy from the expression for linear movement. If the mass of this strip is m, the kinetic energy of the strip is $\frac{1}{2} m r^2 \omega^2$, since $r \omega$ is the velocity of the strip if ω is the angular velocity of the wing at the radius r. The kinetic energy, E_k, of the whole wing is found by adding up the kinetic energies of all the strips which make up the wing and is

$$E_k = \Sigma \tfrac{1}{2} m r^2 \omega^2$$
$$= \tfrac{1}{2} \omega^2 \Sigma m r^2, \tag{1.13}$$

if we assume ω is a constant for all parts of the wing.

Comparing equations (1.12) and (1.13) we see that the moment of inertia of the wing is given by

$$J = \Sigma m r^2, \tag{1.14}$$

a quantity which can be evaluated from the distribution of mass in the wing and the position of the axis of rotation. In other words, J depends on the geometry of the situation. If we call the total mass of the wing M, we may write

$$J = M R^2, \tag{1.15}$$

where R is a quantity known as the radius of gyration of the wing. For a given angular velocity, equation (1.12) shows that the kinetic energy of a system will be greater the larger its moment of inertia. Where a limb is to be moved rapidly, as when a horse is running, it is desirable that its moment of inertia be as low as possible, so that the rapid movement involves a relatively low expenditure of energy. This is achieved by concentrating the muscles (and hence much of the mass) of the limb close to its upper joint.

1.4 Vibrations and Waves OMIT

1.4.1 Free Vibration

Consider a mass resting on a highly polished horizontal surface and attached to one end of a spring, the other end of which is clamped (Figure 1.16). If the mass is pulled away from its resting position and released,

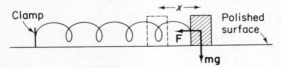

Figure 1.16. Diagram of the system used in the discussion of vibrations. The mass moves on a polished surface under the action of various forces which are discussed in the text

we know that if the spring is of a suitable type the mass will oscillate to and fro. If the displacement of the mass from its initial position is x, the spring exerts a restoring force on the mass of magnitude $-Sx$ where S

(the 'stiffness') is a constant; if friction can be neglected, the acceleration, **a**, of the mass is given by using equation (1.1) :

$$-Sx = ma. \tag{1.16}$$

The minus sign appears because the restoring force in the diagram as we have drawn it is in the direction of decreasing x.

This type of vibrating system, in which the restoring force is proportional to the displacement, undergoes a particular kind of oscillation called simple harmonic motion. A solution of equation (1.16) which gives the displacement x in terms of the time t following the release of the mass at an initial displacement x_0 is

$$x = x_0 \cos (\omega t), \tag{1.17}$$

where $\omega = (S/m)^{1/2}$. A graph of the displacement against time is therefore like a graph of $\cos \theta$ against θ. The significance of the parameter ω can be seen by considering the times at which the displacement again becomes x_0. This occurs every time $\cos (\omega t)$ is unity, i.e. when $\omega t = 2\pi$, 4π, 6π, and so on. The time interval, T, between two successive occasions when the mass is at x_0 is clearly

$$T = \frac{2\pi}{\omega}. \tag{1.18}$$

T is called the period of the vibration while x_0 is known as its amplitude. The frequency, f, of the vibration, or the number of times in one second that the mass returns to x_0, is just $1/T$ and is measured in hertz (Hz). ω is called the angular frequency of the vibration and is related to the frequency f by the equation

$$\omega = 2\pi f. \tag{1.19}$$

A more pictorial idea of the angular frequency may be obtained by considering the relationship between simple harmonic motion and the movement

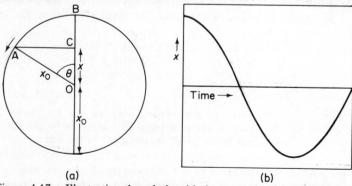

(a)   (b)

Figure 1.17.   Illustrating the relationship between circular and simple harmonic motion

of a point, A, at uniform speed around the circumference of a circle (Figure 1.17a). Let the radius of the circle be x_0 and the angular velocity of the moving point be ω. At a time t after passing the point B the angle AOB in Figure 1.17a is ωt, so that equation (1.17) applies to this situation. The movement along the diameter through O and B of the point C, which is obtained by drawing a line from A perpendicular to OB, is thus simple harmonic in character. The angular frequency ω of the oscillation is simply the total angle swept through by the radius OA in one second.

A graph of displacement against time is shown in Figure 1.17b and represents pictorially the cosine relationship of equation (1.17).

1.4.2 Damped Vibrations

We know that, in practice, the amplitude of the vibrations of the system depicted in Figure 1.16 will gradually decrease as time passes. The most important reasons for this are the frictional force on the weight and the energy dissipated internally by the spring as it stretches and contracts. Oscillations in which the amplitude decreases with the passage of time are called damped vibrations. If the damping forces can be related to the physical properties of the system it is sometimes possible to obtain equations which can be solved to describe the main features of the movement. For example, if the forces impeding the movement of the weight (Figure 1.16) are proportional to the velocity, v, of the weight, the force which is available at any instant to accelerate the weight has a magnitude $(-Sx - Kv)$, where K is a constant, so that equation (1.16) describing the motion now becomes

$$Sx + Kv + ma = 0. \tag{1.20}$$

One approximate solution of this equation is

$$x = x_0 \, e^{-Kt/2m} \cos (\sqrt{S/m - K^2/4m^2} \; t), \tag{1.21}$$

a graph of which is shown in Figure 1.18. Successive maximum amplitudes follow an exponential decay with time, and, from equation (1.21), it is clear that the larger the value K/m the more rapidly will the amplitude decay.

The discussion so far has assumed implicitly that oscillations will occur in the system. Inspection of equation (1.21), however, shows that if the damping coefficient K is large and $K^2/4m^2$ is greater than S/m there is no real solution because the quantity inside the root is negative. In physical terms this simply means that the system will not vibrate; the parameters of the system are such that when the weight is released it will approach its equilibrium position slowly and will not overshoot that position. Such a condition could occur, for example, if the system were immersed in a viscous solution such as treacle (the phenomenon of viscosity is discussed in Chapter 3). The more heavily damped the system (i.e. the greater the value of K) the more slowly will the weight approach its equilibrium position.

The weight will reach its equilibrium position most quickly in the case where $K^2/4m^2$ is equal to S/m. The degree of damping corresponding to this condition is just sufficient to prevent the system vibrating. In this

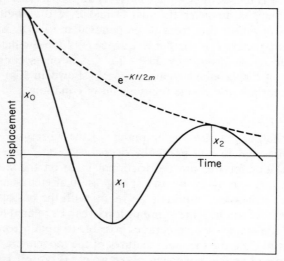

Figure 1.18. Graph of displacement against time for damped vibrations

condition the system is said to be critically damped. Critical damping is of importance in certain measuring instruments (e.g. the moving coil meters discussed in Chapter 6) since it enables measurements to be obtained in the minimum practicable time. It can be shown that the system travels 99 per cent. of the way to its final position in a time of about $6.7 \sqrt{m/S}$ s.

1.4.3 *Forced Vibrations*

By applying an extra periodically varying force to the weight of Figure 1.16 it is possible to produce vibrations of constant amplitude. For obvious reasons these are called forced vibrations. If the applied force is sinusoidal the vibration amplitude will have its largest value at a frequency which is called the *resonant frequency* of the system and is only slightly lower than the frequency of free vibration in a damped system.

Although the frequency of the vibration is the same as that of the applied force, the vibration does not in fact reach its maximum amplitude until some time after the force has passed through its maximum value. We say that there is a *phase difference* between the force and the displacement. The greater the value for K (i.e. the greater the damping) the greater will be the phase lag of displacement behind the force. A graph of the force against the displacement forms a loop (called a hysteresis loop, Figure 1.19).

If the loop is traversed in an anticlockwise direction, work is being done by the force, whereas if the system produces a clockwise loop, work is being done against the force.

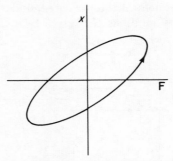

Figure 1.19. A possible relationship between force (F) and displacement (x) in a system executing forced vibrations. The arrow indicates that the loop is traversed in an anticlockwise direction, so that work is being done by the force

In biological systems, examples of damped vibrations are found in vertebrate otoliths and semi–circular canals, while insect flight muscles form part of a system executing forced vibrations. Details of these systems will be found in some of the books listed later.

1.5 Elasticity in Solids
1.5.1 Elastic Moduli

If one attempts to deform a solid, forces which tend to resist the deformation are set up between the atoms or molecules in the solid, giving rise to the property of the solid known as elasticity. The amount of deformation produced by a given applied force depends on the material, and can be predicted from a knowledge of the *elastic modulus* of the material. For a particular material there are three separate elastic moduli, each of which is characteristic of a particular type of deformation. The Young modulus applies when a material is stretched, while the rigidity modulus characterizes deformation brought about by a couple. When a material is deformed under the action of a uniformly applied pressure, the bulk modulus determines the amount of deformation.

All the elastic moduli are defined as the ratio of a stress, which is a force acting on a unit area of the material, to a strain, which is a measure of the relative change in size of the material. Thus, if a piece of wire of length l and cross–section area A is extended by an amount e following the application of a force $\mathbf{F}$ normal to the area, the Young modulus is

$$E = \frac{\text{stress}}{\text{strain}} = \frac{F/A}{e/l}. \tag{1.22}$$

For many materials a range of extensions exists for which the Young modulus remains constant (i.e. the force is directly proportional to the extension). In this range a material is said to obey Hooke's law.

Again, if a material is deformed (or *sheared*) under the action of a couple (Figure 1.20) the rigidity modulus G is given by

$$G = \frac{F/A}{\theta},\qquad(1.23)$$

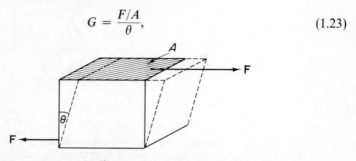

Figure 1.20. Showing the action of a shearing couple, the component forces of which are applied to and lie in the plane of the upper and lower faces of a solid with rectangular faces. The angle θ is a measure of the strain produced by the shearing couple

where the angle θ is taken as a measure of the distortion and A is the area of the face of the solid in which the force lies. Finally, when a solid of original volume V diminishes in volume by an amount v under the action of a pressure P, the bulk modulus is

$$K = \frac{P}{v/V}.\qquad(1.24)$$

The elastic moduli do not necessarily remain constant as a material is progressively deformed. From equations (1.22), (1.23) and (1.24) it can be seen that the units of all the moduli are $N\,m^{-2}$. It can also be seen from these definitions that the more distortion that is produced by a given stress (the more 'elastic' the material in *everyday* language) the smaller is the modulus of the material concerned. Table 1.1 gives values for these elastic moduli for a variety of metals and materials of biological interest.

1.5.2 *Atomic and Molecular Origins of Elasticity*

One knows from experience that there are great differences between the elastic properties of metals and those of rubber. A metal such as steel can only be extended by a small fraction of its original length before it breaks, whereas a rubber band can be stretched to two or three times its original length before it breaks. As can be seen from Table 1.1, the elastic moduli of metals are significantly greater than those of rubber. These differences arise because different molecular phenomena are responsible for the elastic properties of the two types of material.

TABLE 1.1
Elastic moduli of some materials

Material	E (N m^{-2})	G (N m^{-2})	K (N m^{-2})
Aluminium	71×10^9	26×10^9	75×10^9
Steel	210×10^9	83×10^9	168×10^9
Collagen	about 10^9		
Bone	about 10^{10}		
Rubber	about 10^6		

A metal consists of a large number of small crystals, in each of which the atoms are arranged in a regular fashion (Figure 1.21a). When a metal is deformed, the interatomic spacings change (Figure 1.21b) and bring into

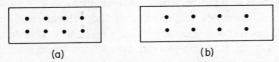

(a)   (b)

Figure 1.21.   Diagrammatic representation of the position of atoms
in a metal (a) before stretching and (b) after stretching

play an elastic restoring force. Rubber consists of an irregular array of long molecules which are cross–linked at various points along their length (Figure 1.22). In the unstretched state the molecules are folded irregularly

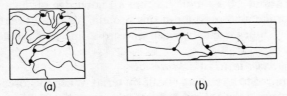

(a)   (b)

Figure 1.22.   Diagrammatic representation of the behaviour of cross–
linked molecules in a material such as rubber when the material
is stretched

and are in a state of continuous agitation because of Brownian motion. When stretching forces are applied, the molecules unfold themselves and become more ordered (Figure 1.22b), while the cross–links prevent the molecules from sliding completely past each other. When the forces are released, the random movements of the molecules tend to destroy the order and hence to shorten the material. Rubbery elasticity, as it is called, is characteristic not only of rubber but of the majority of materials which contain cross–linked long molecules or polymers. This type of elasticity is that which occurs in many biological materials, such as the vertebrate protein elastin which is found in arterial walls and the protein abductin which is used to separate the two valves of a scallop.

The number of cross–links between molecules in a rubbery material affects the amount of deformation that is possible and also the moduli of elasticity. One can see intuitively that if relatively long regions of coiled molecules exist between adjacent cross–links, greater changes in length are possible than if the regions between adjacent cross–links are short. It is possible to derive the following relation between the shear modulus and the average molar mass (M_l) of the molecule between adjacent cross–links:

$$G = \rho \, RT/M_l. \qquad (1.25)$$

In this equation ρ is the concentration of the rubbery material in the solid and takes account of the presence, for example, of water molecules between the long molecules of a rubbery protein. For pure substances such as rubber, the parameter ρ becomes the density of the substance. The symbol R in equation (1.25) is the gas constant, while T is the kelvin temperature.

Let us use equation (1.25) to estimate M_l for the elastic protein resilin, which occurs, for example, in the insect thorax where it serves to reduce the muscular power necessary for flight. The Young modulus for this material is $1.8 \times 10^6 \, \mathrm{N\,m^{-2}}$, so that the rigidity modulus, which is generally about one–third of the Young modulus (Table 1.1), has a value in the region of $6 \times 10^5 \, \mathrm{N\,m^{-2}}$. The value for ρ is about $500 \, \mathrm{kg\,m^{-3}}$, so that from equation (1.25) the average molar mass of the part of a molecule between adjacent cross–links is about $2.0 \, \mathrm{kg\,mol^{-1}}$. From a knowledge of the average molar mass of the amino acid residues in this protein, we can estimate that about twenty–three residues occur between one cross–link and the next.

1.5.3 Energy Stored in Elastic Materials

Work is required to stretch an elastic material. When the forces performing the stretching are released or reduced the material tends to resume its original position, provided the stretching forces were not sufficiently large to disrupt the internal structure of the material. The work done in stretching the material is found by multiplying the elastic restoring force (which, in general, changes during the stretching process) by the distance through which the stretching force moves. This may be conveniently represented by a diagram of force against extension (Figure 1.23a). The work done is the area below the curve, a fact which is easily seen by considering the point on the curve where the restoring force has a magnitude F. On stretching the material by a very small amount, de, the work done against the restoring

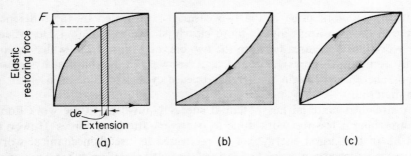

Figure 1.23. Graphs showing the relation between the elastic restoring force and the extension (a) for a material being stretched and (b) when the material is released after stretching. The graphs are combined in (c). The area under graph (a) represents the work done on the material to stretch it, while that under graph (b) represents the work recovered when the material is released. The area between the curves in (c) represents the energy lost in the elastic cycle. Graph (c) is called a hysteresis loop

force is $F\,de$, which is the area of the shaded strip in Figure 1.23a. The total work done is found by adding all such quantities as $F\,de$, a procedure which shows that the work done is the area beneath the curve. Mathematically, the work done, W, is expressed as an integral, so that

$$W = \int_{e_1}^{e_2} F\,de, \qquad (1.26)$$

where the material is stretched from an extension e_1 to an extension e_2. If the force is known as a function of the extension, then equation (1.26) can be used directly to obtain the work done. For example, in a material which obeys Hooke's law the force is proportional to the extension, so that $F = Ce$ where C is a constant and equation (1.26) becomes

$$W = \int_{e_1}^{e_2} C\,e\,de = \tfrac{1}{2}\,C(e_2^2 - e_1^2). \qquad (1.27)$$

More generally, and in particular for elastic materials in biological systems, Hooke's law is not obeyed and the force/extension curve is similar to Figure 1.23a. The form of the graph obtained by plotting the force against the extension when the material is allowed to resume its original position is shown in Figure 1.23b. The curve does not follow that of Figure 1.23a in the reverse direction, but takes a path such that the area beneath it is smaller than that of Figure 1.23a. Thus the work done by the restoring forces during recovery is less than that done by the stretching forces. The difference between the work done in the two cases is represented by the difference between the two areas and is more clearly shown in Figure 1.23c, where the work lost is the area between the two curves. Figure 1.23c, which essentially combines Figures 1.23a and b, is known as a mechanical hysteresis loop. The energy lost in the stretch–release cycle is dissipated as heat in the material.

A stretched material has potential energy equivalent to the work done in stretching it less the work done to overcome frictional forces. However, not all of the stored energy can be recovered as useful mechanical work when the material contracts. The ratio of the work recovered when a material contracts to that which was required to stretch it is called the *resilience* of the material. (The ratio is usually given as a percentage.) The resilience is not a constant of the material, but depends on the duration of the elastic cycle. The maximum value for the resilience of the protein resilin is about 97 per cent., while for abductin it is about 91 per cent.

In our discussion so far we have considered the work which can be obtained from a material which is stretched beyond its resting length. Certain situations exist where work can be recovered following compression of a material. An example is a rubber ball which is compressed if dropped on the ground from a height. The consequent expansion of the rubber causes the ball to rebound, although to a lower height than that from which it was dropped.

The work obtained from stretched or compressed materials is of importance in many biological systems. In vertebrate arteries relaxation of elastin assists the flow of blood, while expansion of abductin provides energy to separate the valves of a scallop. The energy available from deformed resilin is used to provide some of the power required for insect flight and also enables a flea to jump as high as it does. Elastic energy stored in a stretched muscle is used by locusts when they jump.

1.6 Strength of Materials

In general terms the strength of a material is specified in terms of the stress which is just sufficient to break it. The tensile strength is the strength of the material when subjected to stretching forces, while the compressive strength is the corresponding quantity for compressive forces. The cross-sectional area which is used to determine the stress is at right angles to the fracturing force and is measured before the force is applied. In some materials, such as rubber, the cross-sectional area changes before breakage occurs, so that the actual breaking stress has a different value from the tensile strength. The tensile strengths of some metals and biological materials are given in Table 1.2.

The tensile strength of bone is found to be greater than that of the collagen which forms perhaps 50 per cent. (by volume) of the bone. The bone is apparently stronger than the isolated collagen because of the presence of the bone salt, which is in the form of crystals about 20 nm long and is believed to be firmly attached to the collagen.

TABLE 1.2
Tensile strengths of some materials

Material	Tensile strength ($N\,m^{-2}$)
Steel	460×10^6
Aluminium	60×10^6
Collagen	between 50 and 100×10^6
Bone	about 100×10^6
Locust cuticle	about 100×10^6

1.6.1 *Bending*

There are many situations in biology where a system must resist stresses which tend to bend it. One such system is the wing of a bird in flight.

In Figure 1.24 a beam clamped at one end is bent into a circular arc of radius r by applying a couple to its other extremity. The moment, M, of

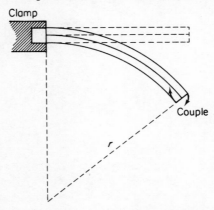

Figure 1.24. A beam bent into a circular arc (radius r) by the action of a couple. The original position of the beam is shown by the dashed outline

this couple is called the bending moment in this situation. Consideration of the figure will show that the upper parts of the beam are stretched while the lower ones are in a state of compression.

By applying a sufficiently large bending moment it is possible to break the beam, and the value of the bending moment, M_F, at the instant of fracture can be predicted if the tensile strength, T, of the material of the beam is known. For instance, if a solid cylindrical rod with a radius, r_1, of cross-section is fractured by bending, then

$$M_F = T \pi r_1^3/4. \tag{1.28}$$

The strength of the rod thus depends on the cube of its thickness in this case. Similar equations show that, for beams of equal weight, a tubular design is stronger than a solid one. The limb bones of vertebrates are hollow and are therefore lighter than solid bones of equal strength. The human femur, for instance, is only about three–quarters of the weight of a solid bone of equivalent strength.

Further Reading

Alexander, R.M. (1968). *Animal Mechanics*. Sidgwick and Jackson, London.

Feltham, P. (1966). *Deformation and Strength of Materials*. Butterworth, London.

Gordon, J.E. (1968). *The New Science of Strong Materials*. Penguin, London.

Hall, I.H. (1968). *Deformation of Solids*. Nelson, London.

Marriott, F.H.C. (1970). *Basic Mathematics for the Biological and Social Sciences*. Pergamon, Oxford.

Morgan, J. (1969). *Introduction to University Physics*, Vol. 1, 2nd ed. Allyn and Bacon, Boston.

Richards, J.A., F.W. Sears, M.R. Wehr and M.W. Zemansky (1963). *Modern University Physics*, pt. 1. Addison–Wesley, Reading, Mass.

Schachman, H.K. (1959). *Ultracentrifugation in Biochemistry*. Academic Press, New York.

Sprackling, M.T. (1970). *The Mechanical Properties of Matter*. English Universities Press, London.

Tricker, R.A.R., and B.J.K. Tricker (1967). *The Science of Movement*. Mills and Boon, London.

PROBLEMS

1.1 When jumping a locust accelerates from rest to $3 \cdot 4$ m s^{-1} in a time of 25 ms. If the angle of take–off is $55°$ to the horizontal and the locust has a mass of 3 g, calculate the force (magnitude and direction) between each hind–leg and the ground. (The fore–limbs are not involved in jumping.)

1.2 The force in problem 1 above is the reaction between the lower end of the locust's tibia and the ground. The power for the jump is provided by a muscle which exerts a force on the upper end of the tibia. The only other force acting on the tibia occurs at the knee joint which divides the tibia in the ratio 1 : 35, the shorter region being nearer the muscle. At a particular moment during take–off the muscle and tibia make angles of $10°$ and $100°$ respectively with the ground. Assuming the tibia to be in equilibrium (and since the mass of the tibia $\ll$ the mass of the locust this assumption incurs little error), calculate the force exerted by the muscle and compare it with the weight of the locust.

1.3 A bush baby accelerates from rest to a take–off velocity of $7 \cdot 0$ m s^{-1} in 1/20 s. The muscles which provide the power for the jump constitute about 9 per cent. of the total weight of the animal. Estimate (a) the work done and (b) the power expended by unit mass of the jumping muscles. Compare these values with those for the locust of problem 1, given that its jumping muscles weigh about 1/20 of the total weight of the insect.

1.4 The mass of a teleost fish's otolith (density 2·9), which moves in a fluid of about the same density of water, is 33 mg. The otolith moves through 0·07 mm under its own weight. If the system is critically damped, estimate the time taken for the otolith to travel 99 per cent. of the way to its final position. (This is a measure of the time in which the fish could be supplied with information about changes in its position.) The coefficient of damping (*K*) is found by experiment to be 15 mN s m^{-1}. How does this compare with the value required for critical damping?

1.5 When the body of a scallop is carefully removed from its shell, the two valves are held open by the action of the elastic protein abductin at the hinge. If one valve is clamped to a bench and the other closed and then released, it will vibrate as a consequence of the elasticity of the abductin. If the ratio of the amplitude of one vibration to the next is 1·05 : 1, calculate the resilience of abductin, assuming that it obeys Hooke's law for compressions employed in the experiment.

1.6 A tree presents an area of 25 m^2 to a wind in which the air moves at 50 km h^{-1}. If all the momentum of the air is destroyed when it strikes the tree and the centre of area of the tree is 8 m above the ground, determine the couple which must be sustained by the root system to keep the tree standing.

What force (in kg wt) would a bulldozer need to exert on the tree at a height of one metre above its base to produce the same effect as the wind? (Density of air = 1·2 kg m^{-3}.)

CHAPTER 2
Temperature and
Heat Transfer

We take it for granted that the temperature of our surroundings should control our lives to a great extent—the clothes we wear, the sports we follow, the effects of the climate and seasons of the country we live in, and so on. The effects of temperature are even more marked for mammals who depend more directly than man on the natural environment, and of course for plant and insect life. In essence the effect of temperature on life is an effect on the chemical reactions (in particular enzymic reactions) which occur in a living organism. Most of these metabolic processes speed up by a factor of two or three times if the temperature is increased by 10°C, so that unless an organism has the means of controlling its internal temperature, its behaviour is dictated directly by the temperature of its environment.

A change in the temperature of a material body, as we shall see, is intimately connected with the amount of 'internal energy' it has gained or lost, so that in order to understand the control of temperature we must understand the various processes by which energy may be transferred from one body to another. In practice, of course, we also require to measure temperature as well as to understand it, so this chapter will also describe some thermometers in common use.

2.1 Temperature and Thermometry

If we place two material bodies together, one of which appears to us to be 'hotter' than the other, then changes occur in the physical properties of both bodies. For instance, if we put a piece of brass from the room on a block of ice, the volume of the brass will decrease, its electrical conductivity will change, and so on, while some of the ice will probably melt (i.e. undergo a change of state).

If, on the other hand, we have two bodies or systems in contact in which physical changes have ceased to occur, we say they have achieved *thermal equilibrium*. All bodies are said to possess a physical property called tem—perature, which determines whether bodies will or will not be in thermal equilibrium; a number of bodies that are all in thermal equilibrium with each other have, by definition, the same temperature. This temperature can be described by a certain number on a temperature scale and, to agree with our subjective experience of heat and cold, higher numbers are assigned to temperatures that appear 'hot' and lower numbers to ones that appear 'cold'.

In physics, to assign a precise number to the temperature of a body, we choose some experimental system (a thermometer) with a physical property (X) whose magnitude varies with temperature. We measure X when the system is in thermal equilibrium with the body of unknown temperature. The value of X is then characteristic of the common temperature of the two systems and can be used to define an appropriate number on the temperature scale (Figure 2.1). Ideally the physical property one chooses

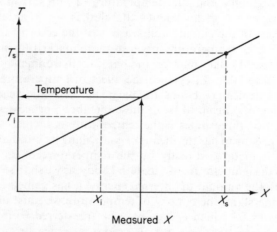

Figure 2.1. The construction of a scale of temperature, T. Values T_s and T_i are defined at the steam and ice points. X is measured at the same points to construct the straight–line graph

(such as the volume of mercury in a mercury–in–glass thermometer) should change in magnitude only with temperature and not with any other factor (e.g. time, atmospheric pressure). Also, X must not have the same value at two different temperatures, otherwise the thermometer would give ambiguous readings.

2.1.1 Temperature Scales

Having chosen a thermometer, it is necessary to construct a scale of temperature. First of all, arbitrary numbers can be assigned to *fixed points* of the scale. These are temperatures characteristic of certain conditions which (a) it is easy to reproduce exactly in the laboratory or (b) can be described in terms of the behaviour of a hypothetical system (such as an ideal gas). For instance, in older definitions of temperature scales, the 'lower fixed point' was the temperature at which ice exists in equilibrium with air–saturated water at standard atmospheric pressure. This 'ice–point' was given the number $0°$ on the Centigrade scale and $32°$ on the Fahrenheit scale. Similarly, the 'upper fixed point' was the temperature of equilibrium between pure water and steam at standard pressure, given the numbers $100°$ and $212°$ respectively on the Centigrade and Fahrenheit scales. The *triple point* is where water, ice and water vapour exist in equilibrium, and since it is more exactly reproducible than the ice point, and is very close to it, it has superseded the latter in accurate definitions of temperature scales. Note that the numbers given to the two fixed points determine the size of the unit of temperature: the Centigrade scale has 100 units between the ice and steam points, while the Fahrenheit scale has 180.

At the moment, our temperature scale has only two numbers, at the fixed points, so methods of interpolation and extrapolation are necessary to find temperatures at other points of the scale. It is here that the choice of thermometer becomes important. Having found the values of X for the thermometer at the fixed points, it is chosen *quite arbitrarily* that other numbers of the temperature scale should be related to values of X by a straight–line graph (Figure 2.1). A measurement of X thus enables us to read off from the graph (or calculate) a corresponding value of temperature. Now X can be any useful property (e.g. the length of mercury in a tube, the resistance of a piece of wire, the volume of a gas) so that, although all thermometers will read the same temperature at a fixed point, it does not follow that the numbers they give for any other temperature will agree. For instance, there is no reason why the volume of a mass of mercury should change with temperature rise in the same way as the resistance of a platinum wire. For this reason, although any thermometer can serve as the basis of a temperature scale, the various scales are not necessarily identical.

It would clearly reduce confusion, in view of what we have just said, to take the temperature scale defined by one particular thermometric property as a standard scale, and calibrate all the other thermometers to agree with it. In fact, this is done by referring to the theoretical properties of an ideal gas. For a fixed mass of such a gas, the product, pV, of pressure and volume is constant at a given temperature (Boyle's law). In theory we could thus use pV as our thermometric property, X, and find the value of pV at, for instance, the ice and steam points by laboratory measurements on an enclosed mass of gas. The appropriate temperature scale would be constructed graphically as in Figure 2.2.

One sees that on producing the straight line backwards, it crosses the temperature axis (i.e. $pV = 0$) at a certain point which on the older Centigrade scale was calculated to be approximately $-273 \cdot 15°$ for an ideal gas. This point ('absolute zero'), at which the product of pressure and volume would become zero for an ideal gas, is used as the lower fixed point of the Ideal Gas Scale and is assigned a temperature of 0 K. The upper fixed point of

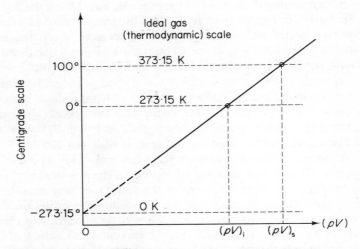

Figure 2.2.   Construction of an Ideal Gas Scale of temperature. The fixed points of this scale are 0 K where $(pV) \to 0$ and 273·16 K at the triple point of water

this scale is the triple point of water, which is given the temperature 273·16 K. The unit of temperature here is called the kelvin (K), after Lord Kelvin, who proposed the adoption of a thermodynamic scale based on the properties of heat engines and independent of any thermometer. The Ideal Gas Scale has been shown to be identical with such a thermodynamic scale, and a temperature measured in kelvins on the scale defined above is called the *thermodynamic temperature, T*.

The interval between absolute zero and the triple point is, by definition, exactly 273·16 K, while on the older Centigrade scale absolute zero was approximately $-273\cdot15°$ and the triple point was approximately $+0\cdot01°$. The degree Centigrade is therefore approximately equal in value to the kelvin. Nowadays, however, the Centigrade scale is no longer used and in its place we have the *Celsius scale* on which a temperature (t) is, by definition, the thermodynamic temperature minus 273·15 K. The Celsius scale is thus the thermodynamic scale, with its origin shifted. The unit of temperature difference is still the kelvin but, for instance, the temperature at absolute zero has a value of exactly $-273\cdot15$ degrees Celsius (°C) and the triple point corresponds to exactly 0·01°C (see Table 2.1).

On the Ideal Gas or Thermodynamic Scale, since $pV = 0$ when the thermodynamic temperature (T) is zero, we have (see Figure 2.2) that $pV \propto T$ for a given mass of gas, or

$$pV = RT, \tag{2.1}$$

if p and V are values for one mole of gas and R is a constant of proportionality called the gas constant. It is evident that the ideal–gas equation (2.1) depends both on Boyle's law and on the definition of the thermodynamic scale of temperature.

TABLE 2.1
Temperatures of the fixed points of various scales

	Thermodynamic scale	Celsius scale	Centigrade scale
Absolute zero	[0 K]	[−273·15°C]	−273·15°
Ice point	273·15 K	0·0°C	[0°]
Triple point	[273·16 K]	[0·01°C]	0·01°
Steam point	373·15 K	100·00°C	[100°]

Note : square brackets enclose values which are assigned exactly, by the definition of the scale.

In practical measurements of temperature it would, of course, be highly inconvenient to try to determine the thermodynamic temperature each time, and so an international committee has laid down the temperatures to be given to a series of additional fixed points, with detailed rules for the types of thermometer to be used in various temperature ranges. The temperatures obtained by this procedure are as near to the thermodynamic scale as measurements allow, and constitute the International Practical Temperature Scale. Examples of the fixed points are the boiling point of oxygen (90·188 K), the freezing point of zinc (692·73 K), the boiling point of sulphur (717·824 K) and the freezing point of gold (1337·58 K).

2.1.2 *Thermometers*

The choice of a thermometer will be limited by the expected magnitude of the temperature to be measured. For example, different types of thermo–meter are needed to measure the temperatures of molten iron and of liquid nitrogen. In general, however, there are certain other criteria by which thermometers should be judged. First, the property X should vary enough for us to be able to detect small temperature differences. Secondly, we want the range covered by the instrument to be as large as practicable. In addition, the thermometer (like all measuring devices) should disturb the system it measures as little as possible; this means it should exchange only a small quantity of heat with its surroundings, and it must therefore have a small 'heat capacity' (see Section 2.2). We also require the thermometer to reach thermal equilibrium with its surroundings quickly, so that we get a quick response to temperature changes and a small time lag. For this, the flow of heat between the thermometer and the system it is measuring should be as efficient as possible. Lastly, it should be easy to calibrate values of X in terms of one of the standard temperature scales.

The best–known thermometer is the *mercury–in–glass* type, in which X is the volume of mercury in a glass bulb. The small changes in volume are made visible as length changes in the mercury thread contained in a narrow capillary drawn out from the bulb. The reliable range of the instrument is from near the freezing point of mercury, $-39°C$, to about $250°C$, where mercury vapour bubbles may form in the thread (the boiling point is $356°C$ at standard pressure). Although the heat capacity is small (i.e. it takes only about 5 J to raise the thermometer's own temperature by 1 K) the glass tends to slow down the heat exchange between the mercury and the exterior, so that a thin–walled bulb is necessary if the thermometer is to respond quickly to changes in temperature. Examples of the errors to which this thermometer is liable are a non–uniform capillary bore (so that equal volume increments produce unequal changes in length of the mercury) and the fact that the stem of the thermometer is usually at a different temperature from the bulb, causing unreliable readings.

The clinical thermometer (Figure 2.3) is a mercury thermometer with a constriction in the capillary arranged so that when the mercury in the bulb

Figure 2.3. The clinical thermometer, showing mercury retained to the right of the constriction in the capillary. The Fahrenheit scale here has normal body temperature marked with an arrowhead. (The usual 0·2° scale divisions are not shown in this diagram.)

contracts after the thermometer is removed from the body, the mercury thread breaks at the constriction. The original (maximum) reading is pre–served by the thread of mercury which remains above the constriction.

An alcohol–in–glass thermometer is used to read lower temperatures, since the alcohol (usually coloured red for visibility) remains liquid between 78 C and -117 C under normal pressure.

Instead of the expansion of a liquid, the expansion of a gas could well be used in constructing a thermometer, although the device would be much more cumbersome. The increase in volume of a gas at constant pressure, or its increase in pressure when constrained to a constant volume, are methods used in the *gas thermometers* which are used to establish the Ideal Gas Temperature Scale. These instruments, which are essentially *standard* thermometers against which others are calibrated, cover a very wide range (c. $-269°C$ to $1600°C$) but are inconvenient for common use.

Electric methods of temperature measurement, on the other hand, have the convenience that a reading can be taken at some distance from the region being measured, and electric devices are often capable of the miniaturization necessary for physiological investigations.

The *platinum–resistance thermometer* uses the fact that the resistance of a metal increases with temperature. Carefully purified platinum wire is used in the standard instrument (Figure 2.4), which has a range from about $-250°C$ to $1300°C$. The wire is wound on an insulating former in such a

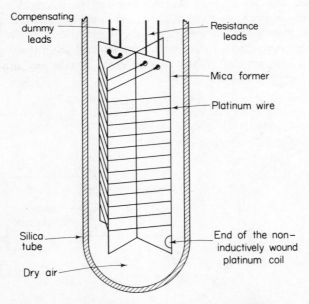

Figure 2.4.   The platinum–resistance thermometer

way that the coil has zero inductance (Chapter 5) and does not pick up stray electrical signals from other apparatus. The dry air in the protective silica tube allows heat to be exchanged between the platinum and the exterior, but the time lag of the instrument may be about 15 seconds.

One could, of course, measure the resistance of the wire at two fixed points and construct a temperature scale as shown in Figure 2.1. This 'platinum scale' differs from the thermodynamic scale, and so it is more usual to calibrate the resistance, R, against fixed points of the International Scale so that the thermometer will give readings on this scale. The variation of resistance with the temperature on this scale in degrees Celsius (t) is found to be

$$R = R_i (1 + At + Bt^2), \tag{2.2}$$

in which R_i is the resistance at the ice point $(t = 0)$ and A and B are constants. By measuring R at two other fixed points, e.g. the steam point $(t = 100°C)$ and the freezing point of zinc $(t = 419\cdot58°C)$, the values of A and B can be derived and tables constructed of t against R.

The resistance, R, is measured with some kind of Wheatstone bridge (Chapter 4) in a way that compensates for variations in the resistance of the leads to the thermometer. In one method (Figure 2.4) identical compensating leads are also taken from the thermometer to the bridge apparatus and connected in such a way that their resistance cancels out that of the real leads, so that only the resistance of the platinum is actually measured.

A more convenient thermometer for general laboratory use is the *thermocouple* (Figure 2.5). When two different metals are joined to form an

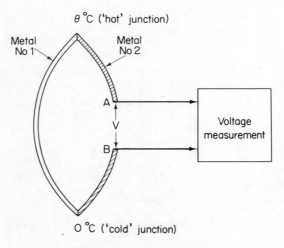

Figure 2.5. The thermocouple

electric circuit and the two junctions between the metals are at different temperatures, an electromotive force (emf) acts so as to cause a current to flow in the circuit. (This is known as the Seebeck effect.) If the circuit is interrupted between A and B as shown, the voltage appearing between these points can be measured, and is a function only of the temperature of the so–called 'hot junction' if the temperature of the other ('cold') junction is kept constant. In practice, the cold junction is immersed in melting ice (0°C) to keep it at a constant temperature. The emf generated by the thermocouple is measured with a microvoltmeter or, ideally, a potentiometer (Chapter 4), which draws no current from the system. (A measurement that depends on current flow depends on the resistance of leads to the thermocouple, and this resistance may vary unpredictably.)

The disadvantage of thermocouple measurements is that, apart from the thermal emfs generated in the thermocouple itself, stray thermal emfs may arise in the leads and parts of the measuring apparatus if they are not all at the same temperature. One advantage of the thermocouple as a thermometer is that the 'hot' junction can be made very small (e.g. two very fine wires, or two very thin metal films, in contact), which additionally means that the system has a small heat capacity and a fast response to changes in temperature. Temperatures can be measured over the range −250°C to about 1800°C, after the variation of emf has been calibrated against a temperature scale.

The calibration graph is a curve, for if E is the emf when the hot junction is at t°C and the cold one at 0 C, it is found that

$$E = at + bt^2, \tag{2.3}$$

and thus the sensitivity (the slope of the curve in microvolts per kelvin) changes with temperature. In the standard thermocouple (for accurate measurements) the two metals are pure platinum and platinum containing 10 per cent. rhodium, while the sensitivity varies from 7.3 μV K^{-1} at 100°C to 12 μV K^{-1} at 1600°C. Simple laboratory measurements can be made

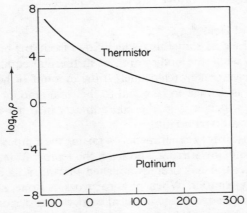

Figure 2.6.   The variation of resistivity, ρ, with Celsius temperature for a thermistor material and for platinum

with a thermocouple of copper and constantan, with a sensitivity of 18μV K^{-1} at $-183°$C and 62μV K^{-1} at $400°$C.

Lastly, one more thermometer which is useful to the biologist is the *thermistor* (i.e. a temperature–sensitive resistor). This is a bead of semi–conducting material (Chapter 6) in which the resistance rapidly *decreases* with increasing temperature, rather than increasing as in a metal (Figure 2.6). The material, a mixture of oxides of, for example, manganese, nickel, cobalt, copper and uranium, is moulded into a bead or disc with two wires attached, then fired in a furnace to produce a hard ceramic substance. For protection it may be mounted in a glass bulb (Figure 2.7) filled with an inert gas to maintain thermal contact, or it may simply be sheathed directly in glass or plastic. The bead can be small enough to be mounted in the end of a hypo–dermic needle for measuring skin temperatures.

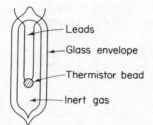

Figure 2.7.   A thermistor thermometer

Because of its small mass, the thermistor has a very low heat capacity and a time lag of as little as 0·1 s in responding to temperature fluctuations. The resistance of the thermistor is usually registered by connecting it in some form of Wheatstone bridge circuit (Chapter 4).

2.2 Measurement of Heat

Heat can be defined as something which flows from one body to another in contact with it while the bodies are not in thermal equilibrium. In the science of heat measurement (calorimetry) the amount of heat transferred from a given body is defined to be equal to the total amount transferred to other bodies with which it is in thermal contact; in other words, heat is assumed to be a conserved quantity.

From nineteenth–century experiments on raising the temperatures of solids by friction and liquids by stirring, and from the temperature rise produced by an electric current, it has been established that heat, as defined above, is a form of energy transfer. When physical work is done, as in stirring a liquid, the energy expended in the physical work is transferred to the liquid and increases what is termed the *internal energy* of the system. The internal energy is the total of the kinetic and potential energies of all the atoms in a given system, and this energy may be increased either by physical work or by allowing the system to absorb heat.

Both heat and work are therefore measured in energy units, such as joules. Historically, however, units of heat were defined before the equivalence of heat and work was recognized, and these units are still in widespread use in chemistry and in domestic heating technology. The definitions of these units are as follows :

One *calorie* is the energy required to raise the temperature of 1 g of water from 14·5°C to 15·5°C (i.e. to raise it by one kelvin at a particular temperature).

One *Kilocalorie* or 'large' calorie is the energy required to heat 1 kg of water from 14·5°C to 15·5°C. Thus 1 kcal = 1 000 cal.

One *British thermal unit* (BTU) is the energy required to heat 1 pound of water from 63°F to 64°F. One *therm* = 10^5 BTU.

Obviously there must exist some conversion factors between these units and the usual units in which energy is measured; such a factor, for historical reasons, is called a 'mechanical equivalent of heat'. The factor can be found experimentally by using mechanical or electrical work (measured in joules) to produce a rise in temperature and then calculating the heat necessary (in heat units) to produce the same effect. The experiment is done in a thermally insulated system from which no energy is lost. One way of doing this is to heat electrically a quantity of water contained in a copper vessel (a calorimeter) which is lagged to prevent heat losses.

If the voltage across the heating element immersed in the water is V and a current of I amperes passes through the element for t seconds, the energy, E, in joules transferred to the system is given by

$$E = V I t. \tag{2.4}$$

Some of the energy is used in heating the heater itself and some in heating the calorimeter and the thermometer it contains, but most is used in raising the temperature of the water. Suppose the calorimeter and all its contents can be expressed as equivalent to a certain mass m of water (in g) and the whole system has its temperature raised through $\Delta\theta$ kelvin by the electrical work. Then the heat necessary (in calories) to produce an identical rise in temperature would be

$$E = m \Delta\theta. \tag{2.5}$$

(This follows from the definition of the calorie, if we make the approximation that it takes one calorie to raise one gramme of water through one kelvin at *any* temperature in the range covered by $\Delta\theta$.)

From the type of experiment we have described, it is thus possible (allowing for the points mentioned above) to obtain two values for the energy transferred from the electrical circuit to the calorimeter system, one in joules and the other in calories, so that the conversion factor can be determined between the two units. Table 2.2 shows the accepted values of this and other factors.

TABLE 2.2
Values of the mechanical equivalents of heat

Heat units		Energy units
1 cal	=	4·186 J
1 BTU	=	1055 J
1 therm	=	$1·055 \times 10^8$ J
3413 BTU	=	1 kWh = $3·6 \times 10^6$ J

The reader may have noticed that, although we can find the energy absorbed by a certain mass of water raised through a certain temperature (using the definitions of heat units and their conversion factors), nothing has yet been said about other substances. It is found that bodies of different substances, although equal in mass, require different amounts of energy to raise their temperatures through the same given interval. For each substance it is therefore possible to measure a value of the *specific heat capacity*, the energy needed to raise 1 kg of the material through a temperature interval of 1 kelvin. The symbol for this value is c_p, where the subscript p denotes a measurement at constant pressure (important with gases). From the definition, c_p is expressed in units of J kg^{-1} K^{-1}. The specific heat capacity may depend on temperature as well as on the particular material; one may, for instance, need less energy to heat a body from 99°C to 100°C than from 0°C to 1°C. Values of c_p are given in Table 2.3.

From the definition of c_p, it follows that if we heat a mass m (in kg) through a temperature interval $\Delta\theta$ kelvin, the energy E required in joules is given by

$$E = m\,c_p\,\Delta\theta, \tag{2.6}$$

if c_p is the average value that applies in the temperature range represented by $\Delta\theta$. For example, suppose we have a type of night–storage heater in which a tank of water, 0·2 m × 0·5 m × 1·0 m in dimensions, is heated overnight to 90°C and cools to 20°C during the following day. The water volume is 0·1 m^3 and its mass is 100 kg. The heat given out during the day is

$$E = 100 \times 4186 \times (90 - 20)$$
$$= 2·93 \times 10^7 \text{ J}$$
$$= 8·13 \text{ kWh} \quad \text{(see Table 2.2).}$$

TABLE 2.3
Values of specific heat capacity, c_p, in J kg^{-1} K^{-1}

Substance	Temperature	c_p
Water	0°C	4218
	15°C	4186
	20°C	4182
	100°C	4216
Ice	−5°C	2060
Aluminium	20 to 100°C	908 (average)
Glass	20 to 100°C	833 (average)
Copper	−223°C	99
	−173°C	254
	20 to 100°C	389 (average)
Mercury	20 to 100°C	138 (average)

This is equivalent to, for example, the energy from an electric radiator working at 1 kW for about 8 hours. If the tank were filled instead with mercury, the mass would be 1360 kg but the value of c_p would be only 138 J kg^{-1} K^{-1}. The heat released would then be

$$E = 1360 \times 138 \times (90 - 20)$$
$$= 1.32 \times 10^7 \text{ J}$$
$$= 3.65 \text{ kWh}.$$

Thus even on a volume–for–volume comparison, where its great density is an advantage, mercury is still less effective than water in storing energy, because the former has such a low specific heat capacity.

The product $(m\,c_p)$ in equation (2.6) is called the *heat capacity* or thermal capacity of the given body of mass m; it takes $(m\,c_p)$ joules to heat that particular body through one kelvin temperature difference. The units of heat capacity are J K^{-1}. Thermal capacity is an important consideration, as we have mentioned earlier, in the use of thermometers, for if a thermo–meter absorbs a given amount of heat when its temperature rises, it takes that heat from the system it is measuring. If the thermal capacity of the system is small, the equilibrium temperature of system–plus–thermometer may be quite altered from the temperature the system would have had before the thermometer was used.

As an example, suppose we have a thermometer with a thermal capacity of 25 J K^{-1} which is at room temperature (20°C), and we place it in 100 cm^3

of water which is initially at 5°C. Let the final temperature reached by the water and thermometer be θ°C. The thermometer, in cooling down from 20°C to the value θ which it finally reads, loses $25(20-\theta)$ joules. The water is raised to θ from 5°C and absorbs $m\,c_p\,(\theta-5)$ joules, where $m\,c_p$ is its thermal capacity. The heat lost by the thermometer is that absorbed by the water, so we have

$$25(20-\theta) = 0{\cdot}1 \times 4200\,(\theta-5),$$

since 100 cm^3 of water has a mass of about 0·1 kg and the specific heat capacity is about 4200 J kg^{-1} K^{-1}. This reduces to

$$445\,\theta = 2600,$$

so

$$\theta = 5{\cdot}84°\text{C}.$$

The temperature measured by this thermometer in these circumstances is thus nearly one kelvin higher than the initial water temperature.

2.3 The Transfer of Heat

There are three natural processes by which heat is transferred from one system to another. The movement of heated material *en masse* from one place to another is termed *convection* and can occur only in fluids (liquids and gases). In *conduction* internal energy is transferred directly from one elementary particle to another (e.g. by collisions or mutual vibrations) and the process can occur in all forms of matter. Energy can also be transmitted from place to place in the form of electromagnetic *radiation* (Chapters 5 and 7), which is able to pass through empty space and thus transfer energy across a vacuum.

2.3.1 *Convection*

Natural convection is a gravitational effect, for when a region of a fluid is heated it generally expands and has a lower density than the rest of the fluid. The general pressure in the fluid exerts an upthrust on the region which is no longer balanced by its weight, so the heated material rises while colder material descends to take its place. In this way circulating currents are established in a fluid which is continuously heated or cooled (e.g. a beaker of water heated at the bottom or a refrigerator cooled at the top, Figure 2.8).

A hot body in otherwise still air loses heat mostly by natural convection, since air is a poor conductor of heat and loss of heat through radiation is negligible except at very high temperatures. The rate of heat loss depends on the difference in temperature, ΔT, between the body and the surrounding air. If the heat *gained* by the body is dq in an infinitesimal time dt, then the heat loss per second can be written as

$$-\frac{dq}{dt} \propto \Delta T^{5/4}, \qquad (2.7)$$

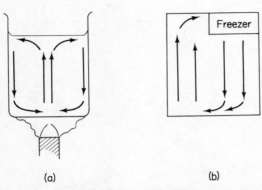

(a)   (b)

Figure 2.8. Convection currents in (a) a beaker of liquid heated from below and (b) a refrigerator cooled at the top

where the exponent 5/4 is an approximate value which has been found by experiment. Since the heat gained by a body equals its thermal capacity ($m\, c_p$) times the rise in temperature (dT for an infinitesimal increase), we have

$$-\frac{m\, c_p\, dT}{dt} \propto \Delta T^{5/4}$$

$$-\frac{dT}{dt} \propto \frac{\Delta T^{5/4}}{m\, c_p}. \qquad (2.8)$$

The cooling rate of a body is therefore inversely proportional to its thermal capacity, while it depends on the 5/4th power of the excess temperature ΔT. (The proportionality factor to be inserted in the equation obviously depends on the shape and surface area of the body concerned.)

In *forced convection*, where air is blown past a body (e.g. with a fan), the exponent in equation (2.8) is found to be 1 instead of 5/4, so the cooling rate is directly proportional to excess temperature, and

$$-\frac{dT}{dt} \propto \frac{\Delta T}{m\, c_p}. \qquad (2.9)$$

The above equation is called 'Newton's Law of Cooling'. In this case the rate of decrease in the excess temperature is proportional to the excess temperature itself, a relationship which gives an exponential decay in excess temperature with increasing time, so that the solution of equation (2.9) is

$$\Delta T = (\Delta T)_{max} \exp\left[-kt/(m\,c_p)\right] \tag{2.10}$$

where k is a constant. The cooling curve is shown in Figure 2.9.

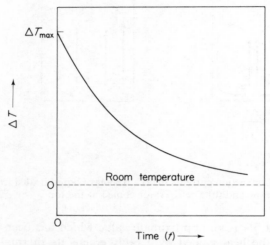

Figure 2.9.   The cooling curve that results from forced convection
(Newton's Law of Cooling) :
$$\Delta T = (\Delta T)_{max} \exp\left[-kt/(m\,c_p)\right]$$

2.3.2 Conduction

In a material where the temperature increases with distance in a particular direction, we say that a *temperature gradient* exists, with a value given at any point by (dT/dx), the temperature rise per unit increase in distance in a specified direction. When a temperature gradient exists, the molecules at the higher temperatures have larger energies of motion or vibration, and there is a net transfer of some of this energy to neighbouring molecules through the material in the direction of decreasing temperature. Heat therefore 'spreads' through the material in directions opposite to the pre-vailing temperature gradients, in a similar way to that in which dye mole-cules in a liquid diffuse outwards when the dye has a concentration gradient. (In fact, the equations describing diffusion and heat conduction are similar in form.)

In a metal electrons are free to move through the molecular structure (Chapter 6), and they constitute a kind of 'electron gas' which permeates the structure and is an efficient conductor of heat. Since free electrons are also responsible, in the main, for conducting charge, it follows that good electrical conductors tend also to be good thermal conductors.

To describe the extent to which a particular material conducts heat, let us consider vithin the material a thin slice of thickness dx and area A (Figure 2.10) in which the temperature at the two faces differs by an amount

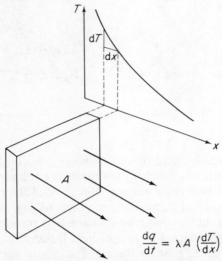

$$\frac{dq}{dt} = \lambda A \left(\frac{dT}{dx}\right)$$

Figure 2.10. Conduction of heat in the x–direction through a thin slice of material. The temperature variation through the material is represented above it

dT. The temperature gradient in the material, in the region we are considering and in the direction x (normal to the plane of the slice), is (dT/dx). If the heat conducted across the slice from one face to the other is dq in a small time dt, then the heat transfer per second is given by

$$\frac{dq}{dt} = \lambda A\left(\frac{dT}{dx}\right), \tag{2.11}$$

in which λ is a constant for the particular material and is called the thermal conductivity. If (dq/dt) is measured in joules per second (i.e. watts), A in square metres and (dT/dx) in kelvins per metre, then λ has the units W m^{-1} K^{-1}, or 'watts per metre kelvin'. Table 2.4 gives values of λ for various solids, liquids and gases. (In practice, conduction in fluids is masked by the greater effects of convection except in special cases, such as fibrous materials, where the movement of air between the fibres is impeded.)

In many cases we are interested in the heat transferred across a wall of

TABLE 2.4
Values of thermal conductivity, λ, in W m^{-1} K^{-1}

Silver	406	Water	0·59
Copper	385		
Aluminium	205	Other liquids	0·1 to 0·2
Steel	50		
Lead	35	Air at 0°C	0·025
Brick, concrete and glass	c. 0·8	H$_2$ at 0°C	0·14
Wood, cork	c. 0·04		

solid material which has a fluid on both sides (e.g. a window pane). In this case the temperature in the bulk of the fluid is not the same as that at the surface of the wall, where there exists a thin layer of fluid which is almost stationary. Heat conducted through the wall has to pass through this fluid 'boundary layer' mainly by conduction, before it is transferred to the bulk fluid by convection currents. The rate of heat transfer from the bulk fluid on one side of the wall to that on the other is clearly not predicted by an equation as simple as (2.11), but can only be described by the use of an empirical factor called the conductance (U) of that particular wall-and-fluid system. If the temperatures of the bulk fluids on either side of the wall are T_1 and T_2, the rate of heat transfer across a given area A is written

$$\frac{\mathrm{d}q}{\mathrm{d}t} = U \, A \, (T_1 - T_2),$$

where U has the dimensions 'watts per square metre kelvin'.

As an example of a calculation of heat conduction, suppose we wish to construct an incubator in which to keep microbial cultures at 37°C, when the lowest temperature of the outer surface of the incubator is expected to be 15°C. We make a cubic box out of polystyrene foam 1 cm (i.e. 0·01 m) thick with sides 50 cm long, and wish to calculate the power of heater to install which will compensate for the maximum expected heat losses through the walls of the box. Neglecting heat losses through the floor of the box, the effective area of the polystyrene walls is $5 \times (0.5)^2 = 1.25$ m^2, while the maximum temperature gradient in the wall would be $(37 - 15)/0.01 = 2200$ K m^{-1}. Since the thermal conductivity of polystyrene is about

$0.04 \text{ W m}^{-1} \text{ K}^{-1}$ the estimated maximum rate of heat loss, using equation (2.11), is

$$\frac{dq}{dt} = 0.04 \times 1.25 \times 2200$$

$$= 110 \text{ W}.$$

We would only need to insert, for instance, an electric heater of 110 W or greater, connected with a suitable thermostat, to maintain the box at the required temperature.

2.3.3 Radiation

In radiation, energy is transmitted in the form of electromagnetic waves, most of the energy being contained in normal circumstances in the infrared part of the spectrum. The energy transfer therefore obeys the ordinary laws of optics, and radiated energy can be reflected, absorbed, transmitted and refracted by suitable materials. The energy radiated from a uniform, small ('point') source obeys the *inverse square law* which applies to all electro–magnetic radiation: the energy per second per unit area received by a surface normal to the rays is inversely proportional to the square of the distance of the surface from the source. The energy per second per unit area (measured in, for example, W m^{-2}) is called the intensity, I, of the radiation, so that if the distance referred to is d, then

$$I \propto \frac{1}{d^2}. \tag{2.12}$$

Depending on the characteristics of their surfaces, most bodies emit (or absorb) radiation with an efficiency which varies with the wavelength of the radiation. For any radiating surface we can define, for a given wave–length, a *monochromatic emissive power*, E_λ. This is the energy radiated per second per unit surface area, per unit wavelength interval of the spectrum. (As we shall see, E_λ depends on the surface temperature.) The energy per second per unit area radiated by a body in the spectral range between the wavelengths of λ and $\lambda + d\lambda$ is then given by $E_\lambda \, d\lambda$.

If the radiation from a body is analysed with a spectrometer, the value of E_λ at a given temperature can be found and plotted against the wavelength, λ, as shown in Figure 2.11. Most surfaces give some irregular curve as shown, which never rises above a limiting curve which represents the monochromatic emissive power of an ideal surface at the same temperature, radiating with maximum efficiency. Since such an ideal emitter of radiation is found also to be a perfect absorber, which reflects no radiation and appears 'black' at all wavelengths, it is called a 'black body'.

In Figure 2.11 the energy per second per square metre emitted by the real surface between the wavelengths shown, λ and $\lambda + d\lambda$, is given, as we

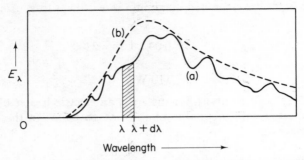

Figure 2.11. The monochromatic emissive power, E_λ, as a function of wavelength for two surfaces at the same temperature: (a) a real surface, (b) an ideal 'black body'

have already said, by $E_\lambda \, d\lambda$. This product, however, corresponds to the shaded area under the graph of E_λ versus λ. Adding such areas over the whole range of the spectrum, we see that the *total* energy per second per unit area, radiated by the surface at all wavelengths, corresponds to the total area under the graph. This quantity is called the *total emissive power*, E, of the surface and can be expressed in joules per second (or watts) per square metre (i.e. W m^{-2}).

As a body gets hotter, it radiates more energy and its monochromatic and total emissive powers increase. By using quantum theory (Chapter 7) it is possible to predict the ways in which the value of E_λ varies both with λ and kelvin temperature, T, for an ideal 'black' body. Curves of E_λ against λ for black–body radiation at various surface temperatures are given in Figure 2.12. The effects on these spectral curves of changing the temperature are to change both their shape and their size. First, we note that the value of E_λ at a given wavelength increases as T increases. Secondly, we see that the total emissive power (the area under a curve) increases very rapidly with T; for instance, the area increases about fivefold in going from 2000 K to 3000 K, a factor of only 1·5 in T. Thirdly, it is apparent that the position of the peak of the curve shifts towards shorter wavelengths (to the left) as temperature increases, so that an increasing fraction of the total radiated energy appears in the visible part of the spectrum.

The second observation, that the total emissive power E increases rapidly with temperature, is described by Stefan's law which states that

$$E = \sigma \, T^4, \tag{2.13}$$

where T is the thermodynamic temperature and σ is the Stefan–Boltzmann constant. If E is to be expressed in watts per square metre (W m^{-2}) and T in kelvin, σ has the value $5{\cdot}67 \times 10^{-8}$ W m^{-2} K^{-4}.

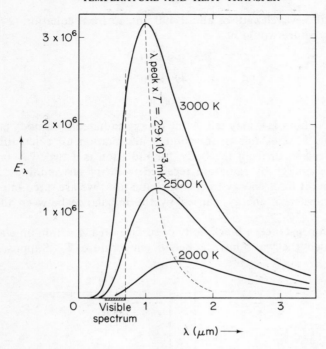

Figure 2.12. Black–body radiation. E_λ is the monochromatic emissive power in W m^{-2} per μm interval in wavelength. Curves are shown for surfaces at three different temperatures. Wavelengths to the right of the visible range of radiation are in the infrared and to the left in the ultraviolet part of the spectrum

The third phenomenon, the shift in wavelength of the peak of the spectral curve, is described by Wien's displacement law, that the wavelength at the peak (λ_p) multiplied by the thermodynamic temperature is a constant. In quantitative terms,

$$\lambda_p T = 2{\cdot}9 \times 10^{-3} \text{ m K} \tag{2.14}$$

so that as T rises, λ_p gets shorter. The colour of a hot body therefore changes as its temperature (and the proportion of energy at the shorter wavelengths) increases. At 600 K the colour of an ideal radiator is pure red; at 1000 K it is pure orange; at 2000 K, yellow; at 6000 K, white; and, at much higher temperatures, blue–white.

One use of Wien's law is to estimate the surface temperature of stars, assuming them to be 'black bodies'. Thus the spectrum of the sun's radiation has a peak at a wavelength of about 490 nm, so from equation (2.14) the surface temperature would be

$$T = \frac{2 \cdot 9 \times 10^{-3}}{490 \times 10^{-9}}$$

$$= 5900 \text{ K.}$$

Although Stefan's law may tell us the energy radiated by a body (strictly only by a 'black' one), this is not enough information for calculating the heat transferred to or from the body, for radiation is a two–way process. A body gains energy by radiation received from its surroundings, at the same time that it loses energy by its own radiation; we are often interested in the *net* transfer of energy, caused by the imbalance between the two processes.

Consider, for instance, a black body of surface area A within an enclosed space of which the walls have a kelvin temperature T_0. Suppose, first,

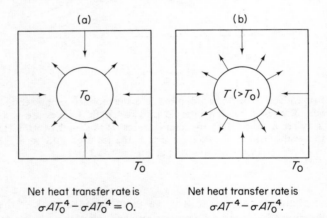

(a)

(b)

Net heat transfer rate is
$\sigma A T_0^4 - \sigma A T_0^4 = 0.$

Net heat transfer rate is
$\sigma A T^4 - \sigma A T_0^4.$

Figure 2.13. The exchange of radiation between a black body and an enclosure at a temperature T_0: (a) thermal equilibrium, (b) body at higher temperature than enclosure

that the body is also at T_0 so that it and the enclosure are in thermal equilibrium (Figure 2.13a). The rate at which energy is radiated from the body is, from Stefan's law, $\sigma A T_0^4$ watts, but since the body remains at the equilibrium temperature T_0, it must also be gaining radiated energy from the enclosure at the identical rate, $\sigma A T_0^4$. If, now, the temperature of the body is changed to a value T by some external means, the energy radiated from it is at a rate of $\sigma A T^4$ watts (Figure 2.13b), while, since the enclosure and the radiation from it remain unchanged, the rate at which energy is *received* by the body is still $\sigma A T_0^4$ watts. The net rate of radiative heat transfer from the black body is therefore given by

$$-\frac{dq}{dt} = \sigma A T^4 - \sigma A T_0^4 \qquad (2.15)$$

when the body is at temperature T and the surroundings at T_0.

2.4 Thermal Effects in Biology

Environmental temperatures on earth are maintained by the energy received as solar radiation (see Chapter 7) which reaches the planet with an average intensity of 1.4 kW m^{-2}. Living organisms themselves produce energy as a by–product of the chemical reactions concerned in their metabolic activity; when an organism is in a steady state, this metabolic energy production is just balanced by the rate at which heat is transferred to the environment. The range of steady–state temperatures within which organisms can survive is fairly limited, which can be understood when we consider the effects of temperature on biological constituents such as proteins, nucleic acids and the fat–like molecules (lipids) of cell membranes.

Proteins, of which enzymes are a particular kind, are large molecules consisting of chains of linked amino acids. In any situation the folded shape (conformation) adopted by the protein chain is the one in which it has a minimum potential energy. A 'native' protein, as found in a healthy organism, has a characteristic conformation which depends on the particular sequence of amino acids along the chain. However, changes in temperature, affecting both the molecule and the solution it is in, can induce the protein chain to alter its conformation, generally by unfolding and becoming more flexible, so that it is no longer able to perform its biological function. The protein has then been thermally 'denatured'. It has been found that many proteins have a temperature of maximum stability in the region of from $10°C$ to $20°C$, and are denatured on approaching both high temperatures (e.g. $40°C$) and low ones ($0°C$).

The rate of a chemical reaction which is catalysed by the action of an enzyme is limited by the amount of enzyme present in its active form. Thus although, in simple theory, the reaction rate should continue to increase exponentially with increasing temperature, in practice the rate rises to a maximum and then falls, as more of the enzyme molecules become denatured at higher temperatures. This means that metabolic activities in an organism have temperatures of maximum efficiency and that radical departures from these values result in malfunction or death.

The effect of increasing temperature on nucleic acids is similar to that on proteins, in that bonds holding the double helices together are ruptured and the two strands of polynucleotide unwind and separate. This 'melting' of the structure generally occurs at higher temperatures than those of the denaturation of proteins.

Membranes and membrane structures within cells (e.g. mitochondria) are damaged by low temperatures ($0°C$ or lower), at which the lipoproteins in the membrane are denatured and lipids are released from the structure. The damage is probably due to changes in the ionic balance of the cell fluid, caused by the slow growth of ice crystals; if cells are rapidly frozen to liquid–nitrogen temperatures, they can often be preserved without gross damage. At high temperatures ($40–50°C$) structural changes or even true melting may occur in the lipid (fatty) part of the membrane, while again the protein may be denatured.

From the above it is obvious that any given organism can only survive within a limited range of internal temperatures. In most organisms the internal temperature (at which metabolic production of energy balances the heat transfer to the environment) rises and falls with the external temperature, although it does not necessarily have the same value. Such organisms are called *poikilotherms* (of many temperatures). Mammals and birds, on the other hand, are able to maintain a constant internal temperature and are called *homeotherms*.

The response of an organism to a change in the external temperature may be threefold: (a) it can make biochemical changes which enable it to function at the new internal temperature; (b) it can attempt to maintain the original internal temperature by controlling its heat transfer; and (c) it can maintain its internal temperature by controlling its energy production. Responses (b) and (c) are called *thermoregulation* and (c) is restricted to homeotherms. Response to the environment may be extended over many generations of a species (when it is termed adaptation), over the life–time of one organism (acclimatization) or over a relatively short period (acclimation).

Examples of the adaptation of poikilotherms to extreme ambient temperatures are the thermophilic algae and bacteria that live in hot springs at temperatures of from $50°C$ to $75°C$ and psychrophilic ones that can grow at $0°C$. Temperature is an important ecological factor in the distribution of plant species, and here, too, we may find conifers that can tolerate $-30°C$ and succulents that tolerate $60°C$. Even at moderate air temperatures, the leaf temperature in a plant could rise to lethal values due to the absorbed solar energy, if most of it were not dissipated by convection. Leaf hairs (e.g. in cacti) improve the efficiency of heat transfer by increasing the effective surface area.

In insects the main process of heat transfer is forced convection (60–80 per cent. of the energy produced in flight by a desert locust is lost in this way), and a coat of hair (in bees) or scales (in moths) seems to be an adaptive response to insulate the insect from convection losses at low ambient temperatures. An interesting example of acclimation is that of ants, which in temperatures of from 20°C to 25°C are found to have no glycerol in the tissues; if the ambient temperature is gradually reduced over a week to between 0°C and 5°C, the ants acquire 10 per cent. by weight of glycerol, a substance known to protect cell membranes from damage on freezing.

Of the higher poikilotherms, aquatic species (fish, reptiles) have internal temperatures very close to the external ones, because the thermal conductivity and high specific heat capacity of water are sufficient to absorb the energy produced by the animal when only a small temperature difference exists. Terrestrial reptiles, however, may have internal temperatures a few kelvin different from that of the air around them. Desert reptiles have developed thermoregulatory responses such as avoiding the sun (a *behavioural* response) and panting at high temperatures, which increases the energy loss due to evaporation of body fluids. (Water absorbs about 2500 J g^{-1} on vaporizing at body temperatures, and thus evaporation is an efficient method of cooling.)

Despite the reponses mentioned above, poikilotherms are essentially at the mercy of the environment, and the homeotherm has a distinct evolutionary advantage in being able to function independently of the external temperature. In addition, the development and functioning of a complex brain is said to be dependent on the maintenance of a constant brain temperature. In fact, homeothermy only refers to the internal organs or 'core' of the body being at a constant temperature, for the peripheral tissues or 'shell' are generally at a lower value. (For example, in man the core is at 37°C while the skin temperature may be 33°C.) Some animals are only 'part–time' homeotherms; for example, the temperature of a humming bird drops at night (nocturnal hypothermia) to conserve the energy which would otherwise be necessary to maintain the day–time value. Hibernating animals undergo seasonal hypothermia.

The metabolic energy production in resting mammals (the so–called basal metabolic rate) is surprisingly constant when divided by their surface area. For instance, from a white mouse weighing 34 g to an elephant weighing 3700 kg, the rate only varies from 42 to 100 watts per square metre. For man the value is about 45 W m^{-2} (i.e. a resting man of surface area 2 m^2 would give out heat at a rate of 90 W). The rate may increase by a factor of up to tenfold under strenuous exercise.

To maintain a constant temperature, metabolic energy must be dissipated to the environment. Heat transfer from the body surface is due principally to convection and radiation, unless there is a layer of hair or clothes through which conduction is necessary first before the other processes can operate. The way heat reaches the surface from the body core is by conduction through the shell and the subcutaneous fat layer and, most importantly, by forced convection of blood flowing from the core to a network of peripheral blood vessels. At normal temperatures, the 'resistance' of the fat layer to the flow of heat is 'short–circuited' by the flow of heat carried by the blood. An additional mechanism of energy loss from the body is evaporative cooling, both by breathing (evaporation occurring from the respiratory tract) and by moisture loss from the skin.

At high ambient temperatures or high metabolic rates, the homeotherm compensates by increasing its rate of energy loss through an increase in the peripheral blood circulation caused by enlarged blood vessels (vasodilation) and by increased evaporative cooling. If there is a hairy coat, the hairs will lie flat to reduce the thickness of the layer of trapped air. In man, vasodilation may increase the effective thermal conductivity through the skin by up to eight times, while he can sweat at 1 kg per hour to achieve a rate of cooling, under optimum conditions, of 700 watts.

At low ambient temperatures the nervous system of the homeotherm acts to reduce heat transfer. Vasoconstriction reduces the flow of blood to the body surface, so that the full insulating properties of the subcutaneous fat are realized. Flow of blood to the body extremities is also curtailed, conserv–ing energy in the body core. Hairs or feathers are raised (piloerection) so that a thicker layer of air, a poor conductor of heat, is trapped near the body surface to provide thermal insulation. If these responses are insufficient to maintain body temperature, energy production is increased—in the short–term by shivering and in long–term acclimation by changes in enzyme activity which lead to increased metabolic activity. As an example of in–creased metabolic activity, wet lambs in a cold wind may generate energy at a rate of up to 290 W m^{-2}. Behavioural responses to cold include nest–building and huddling together to reduce the exposed surface area (e.g. in a flock of sheep).

For accounts of the adaptation of mammals to particular climatic condi–tions and for more details of human thermobiology, we refer the reader to the books listed below.

Further Reading

Burton, A.C., and O.G. Edholm (1955). *Man in a Cold Environment*. Arnold, London.
Crawford, F.H. (1963). *Heat, Thermodynamics and Statistical Physics*. Harcourt, Brace and World, New York.
Edholm, O.G. (1967). *The Biology of Work*. Weidenfeld and Nicolson, London.
National Physical Laboratory (1969). *The International Practical Temperature Scale of* 1968. H.M.S.O., London.
Ramsay, J.A. (1971). *A Guide to Thermodynamics*. Chapman and Hall, London.
Rose, A.H. (Ed.) (1967). *Thermobiology*. Academic Press, New York.
Winslow, C.E.A., and L.P. Herrington (1949). *Temperature and Human Life*. Princeton University Press, Princeton.
Zemansky, M.W. (1968). *Heat and Thermodynamics*, 5th ed. McGraw–Hill, New York.

PROBLEMS

2.1 A certain thermocouple with its cold junction at $0°C$ gives an emf E of 3.60 mV when the hot junction is at $t = 100°C$ and 20.50 mV when the hot junction is at $t = 420°C$. Calculate the constants, a and b, in the equation $E = at + bt^2$, and estimate the sensitivity of the thermocouple, in $\mu V\ K^{-1}$, at temperatures near $0°C$ (i.e. when $bt^2 \ll at$).

2.2 For heating a small greenhouse there is a choice between (a) an electric convector heater rated at 1.0 kW, (b) a storage heater that stores an energy of 25 kWh overnight and releases it evenly over a day–time period of 16 hr, and (c) a gas heater that burns 10 cubic feet of gas per hour from a gas supply with an energy equivalent of 1 therm per 200 cubic feet. Estimate in kW the rate of energy release in (b) and (c), and determine which of the three heaters gives the most power.

2.3 In an experiment on human physiology a nude man is placed in a room with still air at a temperature of $17°C$ and his skin temperature is found to be $33°C$. The air temperature is then raised to $34°C$ and his skin temperature becomes $35°C$. Estimate the ratio between the rates of convective heat loss from the body in the two cases, assuming that the equation for natural convection is applicable.

2.4 A mountain climber has a body surface area of $1.8\ m^2$ and wears fibrous clothing 1 cm thick. He has a skin temperature of $33°C$ while the outer surface of his clothing is at $0°C$. Calculate the total rate of conduction of heat through his clothing, in watts, (a) taking the thermal conductivity of his dry clothing to be 0·04 W m^{-1} K^{-1}, (b) assuming that the clothing is wet through and that the appropriate thermal conductivity is that of water.

2.5 The beetle, *Melanophila acuminata*, lays its eggs in conifers freshly killed by forest fires. Its metathorax has receptors that are especially sensitive to infrared radiation and it can detect and orient towards a fifty–acre area of glowing wood at a distance of 6 km.

Estimate (a) the total rate of energy radiation from such an area, assuming the wood to be glowing orange and behaving as a black body at a temperature of 1000 K; (b) the wavelength corresponding to the peak of the spectral curve at this temperature; (c) the intensity in W m^{-2} of the radiation incident on the beetle at the distance of 6 km. Assume for simplicity that the radiation from the wood is distributed evenly over a hemisphere with its centre at the wood and the beetle at its circumference. (2·47 acres = 10^4 m^2.)

2.6 Solar radiation with an intensity of 1·4 kW m^{-2} falls normally on the upper surface (50 cm^2) of a leaf. Find (a) the total rate at which energy is absorbed by the leaf, assuming none is reflected, and (b) the temperature of the leaf at which it would lose energy at an equal rate by radiative heat transfer, if the average temperature of the environment were 27°C. Assume in (b) that the total surface area (100 cm^2) of the leaf is involved and that all surfaces behave as black bodies.

2.7 In describing the effect of the clothing of man or fur in animals, the *thermal insulation* is often quoted, a value which is the reciprocal of the thermal conductance, U. The fur of the white Arctic fox provides a thermal insulation of about 1·24 K m^2 W^{-1}. At what temperature difference between the inside and outside of the coat would the heat transfer through the fur just equal the animal's basal metabolic rate of, say, 45 W m^{-2}?

CHAPTER 3

Properties of Fluids

The majority of biological systems depend upon fluids for their existence. Most living cells contain liquid, most of which is water, and are supported in a liquid or gaseous environment, the physical properties of which control, to some extent, the behaviour of an organism. Surface tension effects, for example, allow 'water boatmen' to move across the surface of water rather than to sink into it. The combined viscosity and elasticity of synovial fluid make it an excellent lubricant for the mammalian joints in which it is found. The viscosity and density of a fluid environment, together with the shape and other features of an organism, determine the speed with which it can move through the fluid. In this chapter we shall discuss these properties and illustrate their importance with reference to biological systems and certain experimental techniques which are useful in biology.

3.1 Solids, Liquids and Gases

Before discussing in more detail the properties of liquids and gases (collectively known as fluids), it is interesting to compare the molecular configurations and movements of atoms and molecules in solids, liquids and gases. In a solid the atoms are arranged in a regular manner and oscillate about mean positions which remain unaltered within the solid. We say the atoms in a solid have *long–range order*.

In fluids the molecules move about in a random manner. The molecules of liquids tend, on average, to be closer together than those of gases. For this reason the molecules in liquids exert significant forces upon each other when moving between collisions and therefore modify their relative movements. The effect of the intermolecular forces in a liquid is to produce *short–range order*, i.e. regions within the liquid where the molecules are arranged momentarily in a regular fashion. The regions are very small compared with the bulk of the liquid and exist for very short periods of time (about 10^{-11} to 10^{-12} s). In gases the intermolecular forces, apart from those occurring during collisions, can usually be neglected. This means that no order at all exists in the molecules of a gas.

For solids and gases, because of their respective molecular regularity and randomness, it has been possible to formulate theories which predict with reasonable accuracy the behaviour of these phases in terms of their molecular configurations and movements. The short–range molecular order of liquids makes it very difficult to produce a successful theory similar to those of solids and gases.

3.2 Surface Tension

There are many commonly observed phenomena which suggest that, under certain circumstances, the surface of a liquid behaves like a stretched membrane. For example, a hanging drop of liquid has a shape similar to that of a flexible sheet of rubber filled with a liquid. A water droplet suspended in oil of the same density, to neutralize the effects of gravitational forces, is found to assume a spherical shape. For a given volume a sphere has a smaller surface area than any other geometric shape. The surface of the water droplet thus tries to make itself as small as is possible consistent with the constraints upon the system; the water surface is thus behaving rather like a stretched membrane.

As a result of the surface forces the liquid surface has a potential energy. Since any system tends towards a state of minimum potential energy, the surface area contracts until the minimum possible number of molecules remains in the surface of the liquid. The surface area will thus become as

TABLE 3.1
Values of the surface tension for a number
of liquids in contact with air

Liquid	In contact with	Temperature (°C)	Surface tension (mN m^{-1})
Water	air	20 100	72·75 58·9
Benzene	air	20	28·85
Selenium	air	217	92·4
Aniline	air	10	44·1
Ethyl alcohol	air	0	24·05
Glycerol	air	20	63·4
Nitrogen	its own vapour	−183	6·6
Helium	its own vapour	−269	0·012

small as is possible subject to other constraints, such as gravitational forces, which act upon the system. A stable condition will be achieved when equilibrium is reached between the surface forces and the other constraints. Because of the tendency for a liquid surface to contract, we can imagine it to be in a state of tension in which forces act in all directions parallel to the surface. If we imagine a cut to be made in the surface, the *surface tension* (γ) is the force per unit length acting in the liquid surface perpendicular to the cut and tending to pull the cut open. The units of surface tension are thus N m^{-1}. Some examples of the magnitudes of the surface tension for a number of liquids are given in Table 3.1. The temperature at which the measurement was made is also shown in the table, since surface tension decreases as the temperature increases.

3.2.1 *Energy of a Surface*

Because of the molecular forces which exist in the surface of a liquid, the liquid surface has potential energy; that is, it is potentially capable of doing work as it contracts. The potential energy per unit area of surface varies with the liquid. To derive an expression for the potential energy, consider the liquid surface of area A shown in Figure 3.1. Suppose that

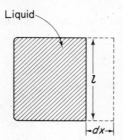

Figure 3.1.   A rectangular liquid surface, originally of area A, increases in area by $dA = l \, dx$

one side of the rectangle is moved through a distance dx to increase the area of the surface by an amount dA. If the length of this side is l, then the magnitude of the force due to surface tension which acts on this side is $l\gamma$, where γ is the surface tension of the liquid. The *mechanical* work done against surface tension forces in moving the side through the distance dx is

$$dW = l\gamma \, dx = \gamma \, dA, \tag{3.1}$$

provided the temperature (and therefore γ) remains constant. The mechanical work necessary to increase the surface area of a liquid by unit area under isothermal conditions is called the free energy (E_F) of the surface. The work done in moving the side of the frame in Figure 3.1 is therefore $dW = E_F \, dA$, so that from equation (3.1)

$$E_F = \gamma. \tag{3.2}$$

The free energy of the surface is therefore equal to the surface tension of the liquid and, of course, has the same dimensions, although it is usual to express free energy in $J \ m^{-2}$ rather than the $N \ m^{-1}$ used for surface tension. (It is left as an exercise for the reader to show that the two expressions are equivalent.)

If the increase in area described above is performed under adiabatic conditions (i.e. so that no heat is allowed to enter or leave the system), the work done on the film is less than the increase in total potential energy of the surface. The difference is provided by a decrease in the internal energy (Chapter 2) of the liquid causing the temperature of the liquid to fall. This follows a general property of physical systems by which, when they are displaced from equilibrium, they tend to set up forces to oppose the displacement. The temperature reduction in this instance increases the surface tension of the liquid, thereby increasing the restoring force acting on the movable side of the rectangle. During the isothermal stretching described earlier, it is clear that the film must absorb heat from the surroundings to keep its temperature constant.

3.2.2 Curved Liquid Surfaces

A curved liquid surface can only exist in equilibrium if an excess pressure

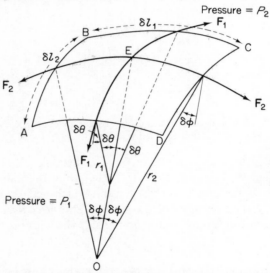

Figure 3.2.   ABCD represents a small area of a curved surface. The principal radii of curvature are r_1 and r_2, while F_1 and F_2 are the surface tension forces which act on the sides of the film. P_1 and P_2 are the pressures on the concave and convex sides respectively of the surface

acts on its concave side to balance the forces due to surface tension effects. Thus, the pressure inside a soap bubble is greater than that outside and prevents the surface from contracting. The excess pressure may be expressed in terms of the surface tension of the liquid and the radii of curvature of the surface. In Figure 3.2 a small area, δA, of a curved liquid surface is shown. In general, it is possible to define two principal radii of curvature, one the maximum and the other the minimum. Arcs with these radii cross each other at right angles in the surface. In Figure 3.2 r_1 and r_2 are the principal radii, while δl_1 and δl_2 represent the lengths of their respective arcs. To a close approximation, the area of the surface considered is equal to $\delta l_1\, \delta l_2$.

Consider the edges AB and DC. A force $\mathbf{F}_2$ due to surface tension acts on each of these sides, where

$$F_2 = \gamma\, \delta l_2. \tag{3.3}$$

If we resolve these forces perpendicular to the line OE which passes through the intersection of the principal arcs, the components we obtain are of magnitude $F_2 \cos \delta\phi$ acting in opposite directions. These forces are therefore in equilibrium. There will, however, be a net force due to surface tension acting parallel to the line OE, and this has a magnitude $F_2 \sin \delta\phi$ for each force. Taking the forces $\mathbf{F}_1$ also into account we find that the total component $\mathbf{F}$ of the forces due to surface tension acting parallel to OE has a magnitude

$$F = 2\gamma\, (\delta l_1\, \delta\theta + \delta l_2\, \delta\phi), \tag{3.4}$$

From Figure 3.2 we have $\delta\phi = \frac{1}{2}(\delta l_1/r_2)$ and $\delta\theta = \frac{1}{2}(\delta l_2/r_1)$, so that on substituting for $\delta\theta$ and $\delta\phi$ and subsequently setting $\delta l_1\, \delta l_2 = \delta A$, equation (3.4) becomes

$$F = \gamma\, \delta A \left(\frac{1}{r_1} + \frac{1}{r_2}\right). \tag{3.5}$$

If the surface is in equilibrium, this force is balanced by the force due to the excess pressure, $P_1 - P_2$, on the concave side of the surface. The magnitude of the force due to the excess pressure is just $(P_1 - P_2)\, \delta A$, so that from equation (3.5)

$$P_1 - P_2 = \gamma\left(\frac{1}{r_1} + \frac{1}{r_2}\right). \tag{3.6}$$

For a spherical surface $r_1 = r_2 = r$ and equation (3.6) becomes

$$P_1 - P_2 = \frac{2\gamma}{r}. \tag{3.7}$$

The excess pressure inside a spherical bubble in air is twice that given by equation (3.7) because the bubble then has two surfaces. Equation (3.7) shows that the smaller the radius of a bubble, the greater will be the excess pressure within it. Thus, if two bubbles of different radii are blown on opposite ends of the same tube by means of suitable side tubes and taps, and then connected together when a further tap is opened, the smaller bubble collapses while inflating the larger one.

3.2.3 Contact between Liquids and Solids

When a liquid is brought into contact with a solid surface, equilibrium is possible only for one particular inclination of the 'free' liquid surface to the solid surface (Figure 3.3). The angle of contact between liquid and

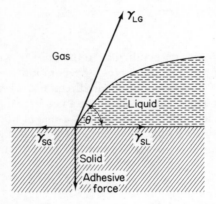

Figure 3.3.   Equilibrium positions of the surfaces of a solid and liquid. θ is the angle of contact. The adhesive force balances the component of γ_{LG} perpendicular to γ_{SG} and γ_{SL}

solid surfaces is the equilibrium angle measured through the liquid. Strictly speaking, we should specify the gas above the liquid when quoting an angle of contact since the gas can influence the angle of contact.

Figure 3.3 represents the relationship between a liquid surface and a solid surface at some instant. We can imagine the forces γ_{LG}, γ_{SL} and γ_{SG} per unit length as shown in the figure, where the subscripts refer to the appropriate interfaces—liquid/gas, solid/liquid and solid/gas. The angle of contact θ can be obtained in terms of the various surface forces by considering the equilibrium at their meeting point. Here we must have

$$\gamma_{LG} \cos \theta + \gamma_{SL} = \gamma_{SG} \qquad (3.8)$$

or

$$\cos \theta = \frac{\gamma_{SG} - \gamma_{SL}}{\gamma_{LG}}. \qquad (3.9)$$

This last equation is known as Young's relation. When a drop of liquid is placed on a solid surface its boundaries will move until the angle θ has the value given by equation (3.9). If the angle of contact is zero or the right-hand side of equation (3.9) is greater than unity, the liquid will continue to spread over the solid surface until it meets a physical barrier. A zero angle of contact is only obtained in practice if the solid has been previously wetted with the liquid, so that a thin liquid film already covers the solid surface. In most cases γ_{SL} is much greater than γ_{SG}. To obtain an angle of contact less than 90°, γ_{SL} has to be negative, which simply means that its direction is 180° displaced from that shown in Figure 3.3.

The angle of contact is extremely sensitive to surface contamination, so that appropriate precautions have to be taken when measuring this parameter. Surface contamination usually lowers the surface tension (γ_{LG}) of the liquid and hence (equation 3.9) the angle of contact θ is reduced. (See also Section 3.2.5.) Some values for the angle of contact of a number of liquid–solid interfaces are given in Table 3.2.

TABLE 3.2
Values for the angle of contact of a number
of liquid–solid interfaces

Interface	Angle of contact	
	degrees	radians
Water–glass	0	0
Most organic liquids–glass	0	0
Water–paraffin wax	105°	1·83
Mercury–glass	140°	2·44
Mercury–amalgamated copper	0	0

Of biological importance is the fact that the angle of contact between water and the waxy surface of a leaf is obtuse. It is therefore impossible for a layer of water to form on the leaf to block its pores. Additives which reduce the angle of contact and thus encourage wetting are sometimes of practical importance. For example, the material used to spray crops for protection against certain diseases usually contains an additive which allows the liquid to wet the leaf surface. Another example occurs in photo-graphy, where a wetting agent is frequently added to the water used to wash a film. This prevents the formation on the film of water drops, which leave marks on the film if left to dry.

3.2.4 *Capillary Rise*

Perhaps one of the most impressive demonstrations of surface tension is the rise of a liquid such as water up a glass capillary tube (Figure 3.4a) for which the angle of contact is zero. Liquids having angles of contact greater than 90°, such as mercury, are depressed in the bore of a capillary (Figure 3.4b). Surface tension forces of the kind shown in Figure 3.3 are

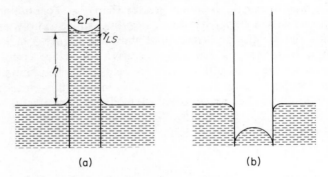

Figure 3.4.   Rise (a) and depression (b) of liquids in capillary tubes

supporting the weight of the column of liquid which stands above the surface of the bulk of the liquid in Figure 3.4a. Let the internal radius of the capillary tube be r and the height above the liquid surface to which the meniscus has risen in the tube be h. Then the perimeter of the circle on which surface tension forces are acting is $2\pi r$, while the volume of liquid supported is $\pi r^2 h$, so that the forces in equilibrium are

$$-2\pi r\,\gamma_{SL} = \pi r^2 h\,\rho\,g. \tag{3.10}$$

Hence, from equation (3.8) we may write

$$2\pi r\,(\gamma_{LG}\cos\theta - \gamma_{SG}) = \pi r^2 h\,\rho\,g. \tag{3.11}$$

In general, γ_{SG} may be neglected when compared with $\gamma_{LG}\cos\theta$, so that equation (3.11) can be simplified to give

$$h = \frac{2\,\gamma_{LG}\cos\theta}{\rho g r}. \tag{3.12}$$

A small correction to equation (3.12) is necessary to allow for the weight of liquid in the meniscus. From equation (3.12) it can be seen that a liquid will rise in the tube if $\cos\theta$ is positive, that is if θ lies between 0 and 90°, whereas it will be depressed if $\cos\theta$ is negative, that is if θ has a value between 90° and 180°.

3.2.5 *Surface Films*

The surface tension of a pure liquid can be markedly affected by the addition of only minute quantities of certain types of impurity. The impurity forms a layer on the surface of the liquid and although to be effective in reducing the surface tension the layer must completely cover the surface, it need only be one molecule thick. If there is not enough of the contaminating material to cover the surface, then the surface tension is only slightly affected. Materials which have this effect on surface tension may be either insoluble or soluble in the liquid concerned. Two types of substance which reduce the surface tension of water are fatty acids, which are insoluble in water, and soaps and detergents, which are soluble. Table 3.3 shows the

TABLE 3.3
Effects of impurities on the surface tension of water

Impurity	Concentration of impurity (%)	Temperature (°C)	Surface tension (mN m^{-1})
None		20	72·75
Sucrose	55	25	75·7
Acetic acid	1	30	68·0
Acetone	5	25	55·5
Phenol	2	20	54·0
Soaps and detergents			c. 27[a]

[a] Minimum value at concentration sufficient to form a monomolecular layer in the surface.

effects of certain impurities on the surface tension of pure water.

Fatty acids, soaps and detergents all have a similar structure. They consist of a 'paraffin' chain with an active chemical group attached to it.

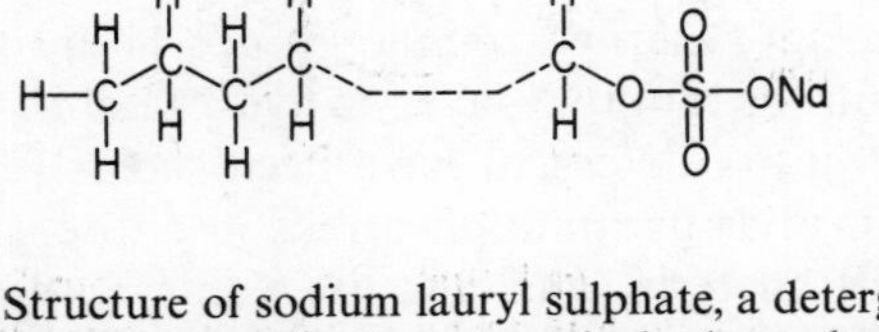

Figure 3.5. Structure of sodium lauryl sulphate, a detergent. In solution, the sodium is removed leaving a negatively charged sulphate group

Figure 3.5 shows the structure of the detergent sodium lauryl sulphate. When in solution, the sodium separates from the rest of the molecule, thus leaving a negatively charged sulphate group. Detergents of this type, which have negatively charged end groups in solution, are called anionic. Cationic detergents also exist, where the end group is positively charged in solution.

The paraffin chains of these substances are *hydrophobic*, that is they are repelled by water molecules. The end groups, on the other hand, are hydro–philic, which means that they have a strong affinity for water. When, for example, soap molecules are placed in water, they tend to aggregate in the surface to form 'islands' called micelles, which have their active groups surrounded by water molecules and their hydrophobic chains as far away as possible from the water molecules (Figure 3.6). If the concentration of

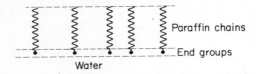

Figure 3.6.   Diagrammatic representation of the arrangement of fatty acid, soap or detergent molecules on a water surface

the soap is sufficiently large, micelles are formed with spherical, lamellar or other arrangements in which the hydrophilic groups are on the outside (Figure 3.7). The structure of the lamellar micelles is similar to that of a cell membrane, which consists of two protein layers separated by a lipid

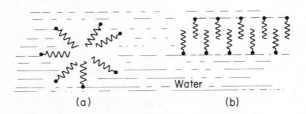

Figure 3.7.   (a) Spherical and (b) lamellar micelles formed by aggregat–ing molecules of a soap, a fatty acid or a detergent

layer. In the cell membrane hydrophilic groups lie in or near the protein layers while the bulk of the lipid layer is hydrophobic. It is interesting to note that certain types of detergent have been successfully used to disrupt cell membranes. This has useful applications in antisepsis, since detergents can be used to kill bacteria, and also in research, where detergents are sometimes used in the isolation of certain cellular components.

3.2.6 *Measurement of Surface Tension*

Many methods are available for the measurement of surface tension. In all of them the apparatus must be extremely clean as surface contamination can dramatically alter the surface tension of a pure liquid. Furthermore, since surface tension varies with temperature, the temperature of the system must remain as constant as possible.

One popular method for the experimental determination of surface tension involves measuring the height to which the liquid in question rises up a capillary tube. Equation (3.12) may then be used to calculate the surface tension. However, it is difficult to clean a narrow capillary tube, so that there is the risk of contamination. Equation (3.12) requires the radius of the capillary at the meniscus of the liquid, a dimension which is difficult to obtain accurately. The angle of contact, too, is an uncertain quantity, and the method is thus usually used for cases where this parameter is zero. Nevertheless, if great accuracy is not required, the capillary tube method can be used to investigate the variation of surface tension with temperature and pressure, and has also been modified by Ferguson to measure surface tensions when only a small amount of liquid is available.

The capillary tube method is an example of a static method, in which the liquid surface remains unchanged throughout the experiment. A method due to Jaeger is a 'dynamic' method, which involves constant renewal of the liquid surface. In Jaeger's method (Figure 3.8a) the end of a tube is

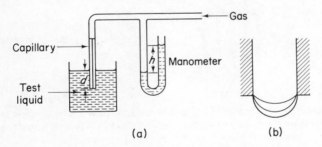

Figure 3.8. (a) Apparatus used in Jaeger's method for the determination of surface tension; (b) is an enlarged view of the submerged orifice showing the decrease in radius of a bubble as the pressure increases. The outermost bubble has the same radius as the orifice

immersed to a depth d in the liquid. A manometer enables the excess pressure in the system to be monitored. Gas is forced through the tube at a suitable rate so that bubbles form slowly at the orifice of the tube. As the pressure increases so the radius of curvature of the bubble decreases, until its radius is equal to that of the orifice (Figure 3.8b). With any further increase in the pressure the bubble becomes unstable and breaks away from the orifice. If h is the maximum difference in heights of the liquid in the two arms of the manometer, the excess pressure within the bubble is $h\rho_1 g - d\rho g$ where ρ_1 and ρ are the densities of the manometer fluid and the liquid under test. This is equal to $2\gamma/r$, where r is the radius of the escaping bubble, so that

$$\gamma = \tfrac{1}{2}(h\rho_1 - d\rho)\, rg. \tag{3.13}$$

From Figure 3.8 it may be seen that the hydrostatic pressure varies over the surface of the bubble, simply because various parts of the bubble are at different depths below the surface. Equation (3.13) is therefore an approximation, and Jaeger's method is useful as a comparative, rather than an absolute, technique.

3.2.7 *Effects of Surface Tension in Biological Systems*

Surface tension plays an important role in a number of biological systems. One of these is the plastron which enables the bug *Aphelocheirus* and certain beetles to obtain their necessary oxygen under water, and thus acts as a kind of gill. The plastron is, in fact, a thin layer of gas which covers most of the body of the organism and which is kept in place by a large number (about 250 million to the square centimetre) of tiny hairs and the surface tension of the water (Figure 3.9).

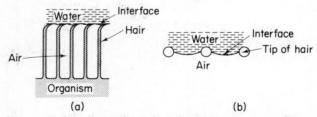

Figure 3.9.   Diagrams of a plastron to show how a film of air is kept in place by the hairs and surface tension effects; (b) is a section at right angles to that of (a) and cutting the tips of the hairs

Surface films are of importance in the lungs of mammals. Such films, which appear to consist of lipoprotein, coat the internal surface of the lung and have been shown to have minimum surface tensions in the region of $2\ \mathrm{mN\ m^{-1}}$, rising to $40\ \mathrm{mN\ m^{-1}}$ as the area of the films is increased. The pressure required to expand the roughly spherical alveoli of the lung is smaller when the lung is deflated than when it is expanded (even when the increase in radii of the alveoli is allowed for). The absence of the surface film causes respiratory difficulties in newly born babies, because the larger surface tension (about $50\ \mathrm{mN\ m^{-1}}$) of the liquid which coats the lungs in this condition is such as to require large pressures to inflate the lung.

3.3 Viscosity of Fluids

3.3.1 *Coefficient of Viscosity*

When a liquid is poured into a vessel we know that it assumes the internal shape of the vessel, although its volume remains constant. Some liquids, such as water, take up the shape rapidly while others, like treacle, can take a considerable time before reaching their final equilibrium states. From this we conclude that liquids offer resistance to flow and have what may be termed internal frictional forces. Gases, too, offer resistance to flow, although the magnitudes of the resisting forces are generally much less than those in liquids.

The gravitational or hydrostatic forces which operate to make a liquid flow usually set up shearing forces within the liquid, since different regions of the liquid move at different velocities. In contrast to solids, fluids do not tend to resume their original shapes when a shearing stress is removed. In fact, a fluid will be deformed continuously as long as a shearing stress is applied. The rate at which the deformation occurs, however, varies from one fluid to another because of the resisting forces which act to oppose the fluid movement. The resisting forces are known as viscous forces and the fluid is said to possess viscosity because of them.

Consider the flow of water along a cylindrical pipe. Provided the rate of flow is sufficiently small, a particular small volume of water will remain at the same distance from the axis of the tube during its passage through the tube. Such flow is said to be *laminar* or *streamline* (Figure 3.10a), and

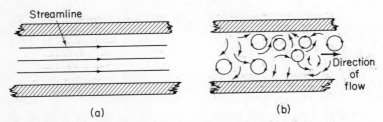

Figure 3.10.   Illustrating (a) laminar or streamline flow and
(b) turbulent flow of water in a cylindrical pipe

the path taken by the small volume is called a streamline. (A streamline is a line such that a tangent to it at any point is in the direction of the velocity of the fluid at that point.) When laminar flow occurs, under steady conditions, the streamlines in the fluid do not change with time. If the rate of flow of water through the cylindrical pipe is increased, a situation is reached where the streamlines become erratic and vortices occur (Figure 3.10b); the flow is then called *turbulent*.

Referring again to the laminar flow depicted by Figure 3.10a, the liquid

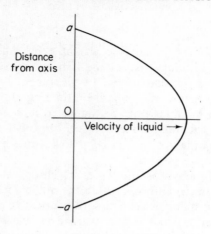

Figure 3.11. Variation of velocity with distance from the axis for a
Newtonian liquid flowing through a tube of radius a

flowing along the axis of the tube moves more rapidly than the liquid near
the wall, as shown in Figure 3.11. In this condition there is a layer of liquid
adsorbed to the wall; this layer remains stationary with respect to the wall
while the rest of the liquid is moving. In Figure 3.12 two layers of liquid,

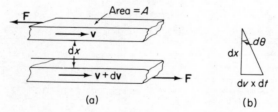

Figure 3.12. (a) Movement of two layers of fluid to illustrate the
meaning of the terms *velocity gradient* and *coefficient of viscosity*.
(b) Diagram to explain the equivalence of velocity gradient and rate
of shear

separated by a distance dx perpendicular to the direction of flow, are shown.
If one layer moves with velocity v and the other with velocity $v + dv$, the
rate of change of velocity with distance is

$$\frac{v + dv - v}{dx} = \frac{dv}{dx}.$$

This is called the *velocity gradient*.

In many contexts one refers to the rate of shear in the fluid rather than to the velocity gradient. The two are, however, equivalent, as may be seen from Figure 3.12b in which only the relative movements of the two layers of fluid in Figure 3.12a are considered. Thus, the faster layer moves with a velocity dv relative to the slower layer. The rate of shear in the fluid is the rate at which the angle $d\theta$ (Figure 3.12b) increases with time and may be expressed as $(d\theta/dt)$. In a small time dt the relative displacement of the two layers is $dv\,dt$, so that provided $d\theta$ is small we have $dx\,d\theta = dv\,dt$, or $(dv/dx) = (d\theta/dt)$.

In many fluids the shearing stress (F/A in Figure 3.12) necessary to set up a given relative movement is found to be proportional to the rate of shear (Newton's law), and we may write

$$\frac{F}{A} = \eta \frac{dv}{dx} = \eta \frac{d\theta}{dt}, \tag{3.14}$$

where η in this case is a constant, known as the *coefficient of viscosity*, or the *dynamic viscosity*, of the fluid. Fluids in which η is a constant independent of the shearing rate are known as Newtonian fluids. The coefficient of viscosity has the units $N\,s\,m^{-2}$, as may be shown from equation (3.14). The poise is the corresponding c.g.s. unit. Some values of the dynamic viscosity of Newtonian fluids are given in Table 3.4. The temperature at

TABLE 3.4
Examples of the dynamic viscosity of Newtonian fluids

Fluid	Temperature (°C)	Coefficient of viscosity (mN s m^{-2})
Hydrogen	0	$8\cdot6 \times 10^{-3}$
Air	0	$1\cdot7 \times 10^{-2}$
Benzene	15	$0\cdot7$
Glycerine	15	$1\cdot5 \times 10^3$
Sucrose	109	$2\cdot8 \times 10^6$
Glucose	22	$9\cdot1 \times 10^{15}$
Water	0	$1\cdot8$
	20	$1\cdot0$
	40	$0\cdot7$
	60	$0\cdot5$
	80	$0\cdot4$
	100	$0\cdot3$

which the measurements were made is recorded in this table because it has a marked effect on viscosity. As the temperature is raised, the coefficient of viscosity falls in the case of liquids but increases for a gas.

3.3.2 *Non–Newtonian Liquids*

There is a large number of liquids for which η in equation (3.14) is dependent on the rate of shear. Such liquids are called non–Newtonian, but can be further classified according to their behaviour when the shear stress is altered. In some liquids, as the shearing stress is increased the rate of shear becomes less than would be expected if η retained its initial value. Such liquids are termed *dilatant* (Figure 3.13). Other liquids (called *thixo-*

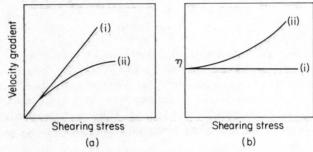

Figure 3.13. Dependence of (a) the velocity gradient and (b) the coefficient of viscosity (η) on the shearing stress for (i) a Newtonian fluid and (ii) a dilatant fluid

tropic) require that the shearing stress decreases with time to maintain a constant rate of shear (Figure 3.14). Many biological liquids (e.g. blood, milk and plant juices) as well as some types of glue and paint are thixotropic.

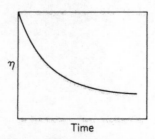

Figure 3.14. Variation of the coefficient of viscosity (η) with time for a thixotropic liquid subjected to a constant shearing stress

The properties of wet sand, which exhibit inverse thixotropy, are probably made use of by burrowing organisms such as the lugworm *Arenicola*.

A material which is inversely thixotropic flows easily under low shearing stress but becomes quite rigid at high values. Possibly the probing, burrowing action of the proboscis of *Arenicola* occurs at a low rate of shear to enable the sand to be moved easily, whereas the movements of the body following the proboscis subject the sand to a high shearing stress, thereby making the burrow walls firm.

The coefficient of viscosity as defined in equation (3.14) only has a meaning for non–Newtonian liquids if the rate of shear is also specified. A measurement made at a given value of the rate of shear is known as the apparent viscosity of the liquid.

3.3.3 Reynolds Number

When an eel swims through water the undulations of its body impart 'backward' momentum to the water. In a system of moving bodies the total momentum is conserved, so that to balance the backward momentum of the water the eel's body acquires 'forward' momentum, and is thus propelled through its liquid environment. The propulsion of the eel depends upon its being able to impart momentum to the surrounding water. Fish, birds and insects are examples of organisms which depend on this type of force for propulsion. Although the viscosity of the fluid plays an important role at the boundary between the organism and the fluid, its effects on propulsion can usually be neglected in these cases. Spermatozoa and microorganisms, on the other hand, depend almost entirely upon viscous forces for their propulsion. Forces derived from momentum effects (often called inertial forces) are negligible by comparison.

Whether the dominant forces acting on a system are viscous or inertial is indicated by the value of a dimensionless parameter called Reynolds number. Thus, large Reynolds numbers mean that inertial forces are important while low Reynolds numbers indicate that viscous forces predominate. For a body moving through a liquid of density ρ and dynamic viscosity η, the Reynolds number (Re) is defined by the relation

$$Re = \frac{\rho \, v \, l}{\eta}, \tag{3.15}$$

in which v is the velocity of the body relative to the stationary fluid and l is a characteristic length (measured in the direction of flow) of the body. The characteristic length of a sphere, for instance, is its radius, while for a cylinder moving parallel to its axis l would be its length. If the cylinder were to move perpendicular to its axis l would be the radius.

Table 3.5 contains Reynolds numbers for a variety of biological systems. Some common non–biological systems are included for purposes of comparison. From the table it is evident that inertial forces are responsible for

TABLE 3.5
Reynolds numbers for a number of systems

System	Reynolds number (order of magnitude)
Submarine; whales and dolphins	10^7 and larger
Aeroplane wings; water snakes; eels	10^6
Birds' wings	10^4 to 10^5
Water beetles; annelid worms	10^3
Insect wings	10^2 to 10^3
Nematodes	1
Beroe comb plates	10^{-1}
Individual cilia and eucaryotic flagella	10^{-3}
Individual bacterial flagella	10^{-5} to 10^{-6}

the propulsion of birds and fish, while the movement of ciliated and flagellated micro–organisms result from viscous forces. For an organism such as a nematode worm, viscous and inertial forces are equally important and a theoretical investigation of this type of system is complex.

The flow of liquid through a pipe is also characterized by a Reynolds number defined by equation (3.15), in which l is a transverse dimension of the pipe. The onset of turbulent flow in a pipe occurs when a particular value of Reynolds number is exceeded. For a cylindrical pipe this value is about 1000.

3.3.4 *Visco–elastic Materials*

Visco–elastic materials are of importance in some biological systems and, as their name implies, they oppose external shearing (or stretching) forces by internal viscous *and* elastic forces. To a first approximation, the viscous and elastic forces can be considered separately, and the internal stress will be the sum of the viscous and elastic stresses. Thus, if an external shearing force **F** is applied tangentially to an area A of a visco–elastic material, we can write

$$\frac{F}{A} = G\theta + \eta \frac{\mathrm{d}\theta}{\mathrm{d}t} \qquad (3.16)$$

where G and η are the rigidity modulus and coefficient of viscosity res–

pectively of the material (see equations 1.23 and 3.14).

A solution of this equation is

$$\theta = (1 - e^{-\frac{G}{\eta}t}) \frac{F}{AG} \qquad (3.17)$$

which shows (Figure 3.15) that for large values of t the strain θ approaches F/AG, which is the equilibrium strain for a purely elastic material of rigidity

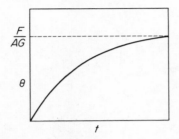

Figure 3.15. Variation of the strain (θ) with time (t) for a visco–elastic material. Equation (3.17) describes the form of this curve

modulus G subject to a shear stress F/A. The ratio η/G has the dimensions of time and is known as the retardation time for the material.

If viscous flow occurs in a visco–elastic material, the elastic forces cannot restore the original shape of the material when the shearing stress is released. Since the amount of deformation depends upon the time for which the material is strained, so also will the extent to which the material recovers following the release of the applied stress. A material strained for a short time will suffer little viscous flow, and almost complete recovery is possible when the deforming forces are released. When a material is strained for a long time, on the other hand, much flow may occur and recovery cannot be complete. In practical terms a visco–elastic material can respond like a rubbery solid to a sharp blow and like a viscous liquid if the applied stress is prolonged.

The mesogloea of the sea–anemone is a visco–elastic material and its properties allow the organism both to withstand sharp knocks and to expand itself enormously at a slow rate without much increase in the internal pressure.

Another important visco–elastic material is synovial fluid which lubricates the synovial joints of mammals. Because of the visco–elastic properties of synovial fluid it is very difficult to remove the fluid completely from between two surfaces by squeezing the surfaces together. For a lubricant this is clearly a very useful property.

3.3.5 *Measurement of Liquid Viscosity*

One of the most convenient and satisfactory instruments with which to measure the viscosity of a Newtonian fluid is the Ostwald viscometer, one form of which is shown in Figure 3.16. An Ostwald viscometer is usually made of glass and has a U–shape with two (or more) reservoirs. To use the

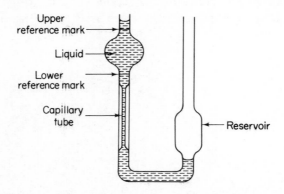

Figure 3.16.   Diagram of one type of Ostwald viscometer

instrument a specific volume of liquid (usually 10 or 20 cm^3) is poured into the right–hand arm of the U, which is made of tubing with a fairly wide bore. The liquid is then sucked into the left–hand arm of the U until the meniscus rises above the upper reference mark. The liquid is next allowed to flow through the capillary tube under the influence of gravity, and the time taken for the meniscus to fall from the upper to the lower reference mark is recorded.

In the Ostwald viscometer liquid is forced to flow through a cylindrical tube because of a pressure difference between the ends of the tube. If the pressure difference is p and a volume V of liquid flows through a tube (length l, radius r) in a time t, it can be shown that

$$\frac{V}{t} = \frac{\pi p a^4}{8 \eta l},$$

(3.18)

a relationship known as Poiseuille's equation. In this case V is the volume enclosed between the two reference marks of the viscometer. From Figure 3.16 it may be seen that the pressure (due to the 'head' of liquid) which drives the liquid through the capillary changes as the liquid flows. For this case it can be shown that an average value $\bar{p} = \bar{h} \rho g$ can be used for p in equation (3.18), where ρ is the density of the liquid and $\bar{h}$ is the average height of the liquid meniscus in the left arm of the U above that in the right arm; $\bar{h}$ is sometimes called the average head of the liquid. It is difficult to determine accurately the values of $\bar{h}$ and l for a particular instrument, and so the Ostwald viscometer is used to compare viscosities rather than for making absolute measurements. If t_1 and t_2 are the times taken with two different liquids for the meniscus to travel from the upper to the lower reference mark, then by using equation (3.18) and the expression of $\bar{p}$ in the two cases, and subsequently dividing, we obtain

$$\frac{\eta_1}{\eta_2} = \frac{\rho_1 \, t_1}{\rho_2 \, t_2}. \tag{3.19}$$

The ratio η/ρ is known as the kinematic viscosity, v, of a liquid and so, from equation (3.19),

$$\frac{v_1}{v_2} = \frac{t_1}{t_2}. \tag{3.20}$$

This implies that the time of flow, t, measured with the viscometer is directly proportional to the kinematic viscosity of the liquid. However, in deriving equation (3.20) no account is taken of the kinetic energy acquired by the liquid in flowing through the capillary; it is thus more exact to express the kinematic viscosity as

$$v = At - \frac{B}{t}, \tag{3.21}$$

where A and B are constants. It is therefore usual to plot a calibration curve of γ against t for the instrument so that the kinematic viscosity of a liquid can be directly read from the curve after its time of flow through the viscometer has been measured.

Because of the small physical dimensions of the Ostwald viscometer it is easy to immerse it in a thermostatically controlled water–bath, and hence to study the effects of temperature on the viscosity of a liquid.

The Ostwald viscometer is useful for measuring the viscosity of fairly mobile fluids, and has been used to determine the viscosities of many solutions of biological interest. For very viscous Newtonian fluids the Redwood viscometer is more convenient. This, like the Ostwald instrument, does not give absolute values directly. It consists essentially of an open cup with a hole in its base. The experimental procedure is to fill the cup up to a particular mark and to time the discharge of a given volume of liquid. The viscosity may then be read from a calibration graph.

In both the Ostwald and Redwood viscometers, the rate of shear (or velocity gradient) changes as the flow proceeds. For Newtonian liquids this is of no consequence, but for non–Newtonian liquids it is essential to make measurements with an instrument in which the rate of shear is constant and known. Examples of such devices are the Couette rotating cylinder and the cone–and–plate viscometers. The latter is illustrated in Figure 3.17. The cone, which has a large angle at its apex, is fixed, while the plate can be rotated below it at various angular velocities about the axis of the cone.

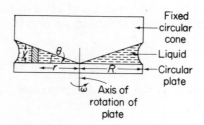

Figure 3.17.   Illustrating the principle of a cone–and–plate viscometer

The liquid under test is located between the cone and the plate, and will consequently be subjected to a shearing stress when the plate rotates. To evaluate the rate of shear, consider an element of liquid at a distance r from the axis of rotation and let its height parallel to the axis of the cone be y. That part of the liquid in contact with the cone is stationary whereas the liquid in contact with the plate moves with a velocity $r\omega$ (Chapter 1). The rate of shear or velocity gradient is thus $r\omega/y$. Now $y/r = \tan\theta$ (Figure 3.17), so the velocity gradient is $\omega/\tan\theta$ and is independent of the distance of the element from the axis of rotation. Thus from a knowledge of the rate of rotation of the plate and the angle θ, it is possible to evaluate the rate of shear experienced by the liquid. From the couple (or *torque*) required to rotate the plate at a given angular frequency the apparent viscosity can be calculated.

3.4 Diffusion

Most of us are familiar with experiments of the type where water is poured carefully onto a coloured liquid such as iodine in a beaker so that a sharp boundary is formed between the two liquids. When some time has elapsed it is obvious from the coloration that some of the iodine has become mixed with the water, even though no shaking has occurred to disturb the boundary. This type of mixing arises because of the random molecular movements in the liquids and is called diffusion. Diffusion also occurs in gases and can occur to a limited extent at the surfaces of solids.

The possibility of diffusion arises in biological cells where many different species of molecule are found in a more–or–less liquid cytoplasm. It is of interest to know how rapidly the molecules can diffuse in the cytoplasm to find out whether this simple process can account for the association of molecules which enables a cell to live and multiply. Such knowledge can only come from a quantitative consideration of the diffusion process. The diffusion process also occurs in the Ouchterlony technique, in which an agar gel is used which contains wells for the introduction of antibodies and antigens. The antibodies and antigens diffuse into the gel and, when they meet, produce a precipitate. The positions of the precipitates can be used to obtain information about the antibodies and antigens, but, once again, knowledge of the rate of diffusion of the molecules is necessary.

Since molecular movement is random and there is therefore no preferred direction for molecular motion, we might wonder why it is that a net movement of molecules occurs from regions of higher concentration to those of lower concentration. This can be explained if we consider the movement of molecules between two equal adjacent volumes in a fluid, and suppose that the number of molecules of a given species is greater in one volume than the other. Although the movement of the molecules is random we may say that in a given time a certain fraction of the molecules from one volume will be transferred across the boundary into the other. Since the same *fraction* will move out of each volume, more molecules will move from the more highly populated volume than from the other. There will thus be a net transfer of molecules from the region of higher concentration to the other.

Consider a plane of area A (Figure 3.18) in a fluid and suppose that the

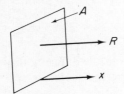

Figure 3.18. Illustrating the diffusion of material in the x-direction across an area A perpendicular to that direction. R is the rate of transfer of material across the area

rate of transfer of diffusing material across the area in the direction x is R. It has been found that

$$R = -DA \frac{\partial c}{\partial x}, \tag{3.22}$$

where $\partial c / \partial x$ is the concentration gradient measured in a direction perpendicular to the plane and D is a molecular parameter called the diffusion coefficient which depends upon the solvent, concentration of solute and the temperature. In some cases D remains sensibly independent of the concentration but in others, particularly where long chain–like molecules are involved, D varies markedly with the concentration. The negative sign of equation (3.22) is inserted because the net transfer of material occurs in the direction of decreasing concentration. Thus R is positive (i.e. in the direction of increasing x) when $\partial c / \partial x$ is negative, and if D is to be given positive values, the minus sign is necessary in the equation.

Equation (3.22) is known as Fick's first law of diffusion. His second law, which is derived from the first, has the form for one–dimensional diffusion of

$$\frac{\partial c}{\partial t} = D \frac{\partial^2 c}{\partial x^2}. \tag{3.23}$$

For diffusion in three dimensions (x, y, z) the equation becomes

$$\frac{\partial c}{\partial t} = D\left(\frac{\partial^2 c}{\partial x^2} + \frac{\partial^2 c}{\partial y^2} + \frac{\partial^2 c}{\partial z^2}\right). \tag{3.24}$$

The solutions of equations (3.22) and (3.24) depend upon the initial distribution of material. For example, if a mass m of material is deposited in a plane at a position $x = 0$ and time $t = 0$, then, for one–dimensional diffusion, it can be shown that the concentration, c, at a point x and time t is given by

$$c = \frac{m}{2(\pi D t)^{1/2}} \exp\left(-\frac{x^2}{4Dt}\right). \tag{3.25}$$

In practice, the initial position of the material is not all in an infinitely thin plane but distributed over a finite volume of space, and the exact solution of the diffusion equations then leads to more complex relationships than equation (3.25).

3.4.1 Relation of Viscosity to Diffusion

Let us consider for the moment a single solute molecule diffusing through a liquid. By virtue of its movement, albeit random, the molecule will experience frictional retarding forces of a viscous nature, similar to those discussed in Section 3.3. The retarding force, $\mathbf{F}$, is directly proportional to the velocity, $\mathbf{v}$, of the molecule and may be written, in general, as

$$F = f\, v, \tag{3.26}$$

where f is called the frictional coefficient.

It can be shown that

$$f = K\eta\, l, \tag{3.27}$$

where η is the viscosity of the liquid in which the molecule is diffusing, l is a linear dimension of the molecule and K is a constant which depends on the shape of the molecule. It follows that the frictional coefficient f also depends on molecular shape. For spherical molecules of radius l, K has the value 6π, leading to the force equation

$$F = 6\pi\, l\, \eta\, v, \tag{3.28}$$

which is known as Stokes' law. The only other shape which is amenable to mathematical calculation of the frictional coefficient is an ellipsoid, and this is the shape assumed in practice for many molecules, particularly those of high molar mass ('molecular weight').

By considering the average distance moved by a molecule in a given time it is possible to show that the diffusion and frictional coefficients are related by the equation

$$D = \frac{kT}{f}, \tag{3.29}$$

where k is the Boltzmann constant and T is the temperature on the kelvin scale. We can use equation (3.29) to get a rough idea of how the diffusion coefficient varies with molar mass, for, if the diffusing molecules are spherical and of radius l, the molar mass M is

$$M = \tfrac{4}{3}\pi\, l^3\, \rho\, N_A, \tag{3.30}$$

where ρ is the density of the molecule and N_A the Avogadro constant. From equation (3.29) and using $f = 6\pi\, \eta\, l$, and equation (3.30), one obtains

$$D = \frac{kT}{6\pi\eta} \sqrt[3]{\frac{4\pi\, \rho\, N_A}{3M}},$$

from which it is apparent that the diffusion coefficient for spherical molecules should be inversely proportional to the cube root of the molar mass. A similar result can reasonably be expected for non–spherical molecules, so that the diffusion coefficient is not expected to vary rapidly with changing molar mass. This is borne out by experiment, as shown by the diffusion coefficients listed in Table 3.6. Since diffusion coefficients vary according to the solvent and temperature, they are usually quoted for water at a temperature of 20°C. Where concentration of the solute affects the diffusion coefficient the quoted value is obtained by extrapolating experimental data to zero concentration. It is important to realize that the diffusion coefficient can only be used to calculate the molar mass if the molecular shape is known. In practice, the molar mass of a molecule is usually measured by other methods and the diffusion coefficient is used to obtain information about the molecular shape.

3.4.2 *Measurement of Diffusion Coefficients*

A method of measurement which has been used with varying degrees of success is the free diffusion technique, in which it is arranged that the pure solvent is initially separated by a sharp boundary from the solution. On the solution side of the boundary, the concentration of diffusing molecules is finite and constant (say C_0), while on the other it is zero. Initially, then,

TABLE 3.6
Diffusion coefficients for molecules diffusing through water

Diffusing molecule	Approximate shape	Molar mass (kg)	Diffusion coefficient $(m^2 s^{-1} \times 10^{11})$
Glycine	small sphere	0·075	95
Arginine	small sphere	0·174	58
Cytochrome C	globular	13	10·1
Lysozyme	globular	14·1	10·4
Pepsin	globular	35	9·0
Haemoglobin	globular	68	6·9
Collagen	long chain	345	0·69
Myosin	long chain	493	1·16
Deoxyribonucleic acid	long chain	6 000	0·13
Tobacco mosaic virus	rods	50 000	0·3

the concentration and concentration gradient will appear as shown by the curves labelled $t = 0$ in Figure 3.19.

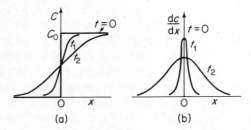

Figure 3.19.   Showing the changes in (a) concentration and (b) concentration gradient with distance (x) and time (t) for a diffusing system in which a sharp boundary existed at $t = 0$ between a solution (concentration C_0) and the solvent

The solution of the diffusion equations (3.22) and (3.23) appropriate to this situation is

$$C = \frac{C_0}{2}\left[1 - \frac{2}{\sqrt{\pi}} \int_0^{\frac{x}{\sqrt{4Dt}}} e^{-y^2} dy \right] \qquad (3.31)$$

and the diffusion gradient, dc/dx, is given by

$$\frac{dc}{dx} = \frac{C_0}{\sqrt{4\pi Dt}}\, e^{-x^2/4Dt} \tag{3.32}$$

The integral in equation (3.31) is the probability integral and is tabulated in many handbooks for various values of the upper limit $(x/\sqrt{4Dt})$ of the integral.

In an experiment, the diffusion process is often followed by measuring $d\mu/dx$, the gradient of refractive index in the liquid. This quantity is proportional to dc/dx, so that equation (3.32) can be used to obtain the diffusion coefficient. Other methods of measurement yield the concentration at a particular position in space and time, and equation (3.31) is then used to calculate D.

The act of setting up the initial sharp boundary is a difficult procedure, in that the movement necessary to bring the liquids together may introduce mechanical disturbances of the liquids. For this reason, the experiment is sometimes performed through a porous membrane, as shown in Figure 3.20.

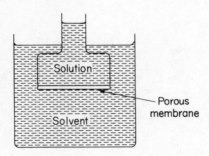

Figure 3.20. Showing the arrangement of vessels and porous membrane in an apparatus used to measure diffusion coefficients

In this case it is necessary to consider the frictional effects of molecules moving through the narrow pores of the membrane. In practice, such an apparatus is calibrated using liquids with well–established values of the diffusion coefficient.

Some of the experimental difficulties associated with the methods described above are overcome in a new technique which involves the scattering of light from a laser. When coherent radiation (Chapter 9) is scattered by a system of randomly moving particles, the intensity of the scattered beam is observed to fluctuate. The fluctuations arise because the frequency of the incident beam is changed slightly on being scattered by a moving molecule (this is an example of the Doppler effect) to an extent which depends on the velocity of the molecule. The consequent superposition of beams scattered by different molecules produces interference effects and a change in the overall intensity of the beam. The fluctuations are related to the relative movements of the molecules and may therefore be used to determine diffusion coefficients. Although the method is not yet widely used, it is capable of producing results more quickly and more accurately than previous techniques.

In this chapter we have discussed the physical properties of fluids which are of interest and importance in biological systems. Some of the books below describe how knowledge of these properties can provide information about biological macromolecules.

Further Reading

Alexander, R.M. (1968). *Animal Mechanics*. Sidgwick and Jackson, London.

Crank, J. (1964). *The Mathematics of Diffusion*. Oxford University Press, London.

Dinsdale, A., and F. Moore (1962). *Viscosity and Its Measurement*. Chapman and Hall, London.

Flowers, B.H., and E. Mendoza (1970). *Properties of Matter*. Wiley, London.

Morgan, J. (1969). *Introduction to University Physics*, Vol. 1, 2nd ed. Allyn and Bacon, Boston.

Sprackling, M.T. (1970). *The Mechanical Properties of Matter*. English Universities Press, London.

Tabor, D. (1969). *Gases, Liquids and Solids*. Penguin, London.

Tanford, C. (1961). *Physical Chemistry of Macromolecules*. Wiley, New York.

PROBLEMS

3.1 In an insect plastron (see Figure 3.9) the lateral separation of adjacent hairs is $0{\cdot}6$ μm and each hair has a diameter of $0{\cdot}2$ μm. The liquid surface makes contact with each hair on the line joining the centres of neighbouring hairs (Figure 3.9b). If the angle of contact between water and the material of the hair is $110°$, calculate the excess pressure which exists across the surface and specify on which side the pressure will be the greater. (Assume the water surface to be cylindrical and take the surface tension of water as 72 mN m^{-1}.)

3.2 An echiuroid worm feeds by passing a current of water through a mucous net 10 μm thick containing circular holes each of area 1 μm^2. The area of the net is 2000 mm^2 and the overall area of mesh is equal to the overall area of void. If $1{\cdot}2$ dm^3 of water flow through the net in an hour, calculate (a) the pressure difference which exists across the net and (b) the power necessary to produce this rate of flow. (Take the viscosity of water as 1 mN s m^{-2}.)

3.3 A beating cilium uses up energy at each point along its length. The energy is supplied by hydrolysis of adenosine triphosphate (ATP) which is used up at the rate of 5 mol s^{-1} m^{-3}. Assume that the cilium is a hollow cylinder which is open at the end where it joins the cell and closed at the distal end. If the cilium is 50 μm long and has a cross–sectional area of $0{\cdot}03$ μm^2, calculate (a) the minimum concentration which must be maintained at the proximal end to keep the entire cilium in motion and (b) the amount of ATP in the cilium. (Assume that ATP is made available by diffusion in one dimension. Take the diffusion coefficient of ATP as $2{\cdot}5 \times 10^{-10}$ m^2 s^{-1}.)

3.4 It is frequently convenient to construct scale models of biological systems to investigate certain aspects of their behaviour. The model should have the same Reynolds number as the real system if conclusions drawn from the model are to be correct for the living organism. A model of a bundle of bacterial flagella is made using a cylinder 0·5 cm in diameter. The lateral velocity of the cylinder is 2 cm s^{-1}. If the Reynolds number for the flagellum *in vivo* is 10^{-3} determine the kinematic viscosity of the fluid which should be used for the model experiment. What would be a suitable fluid for the experiment?

CHAPTER 4

Electricity and Magnetism

The importance of some understanding of electricity does not need to be stressed in an age when the use of electric devices and electronic instrumentation is increasing, both in the laboratory and in the everyday world. At the same time, electric forces are intimately involved in the structure of matter and its behaviour at the molecular level. In order to understand, for instance, the process of nerve conduction or the way in which an antibody molecule binds itself to the antigen, or the effect of a protein molecule on polarized light, some insight is necessary into the nature of electric fields.

The word *field*, as we have just used it, has a specialized meaning and will occur frequently in the next few chapters. If, in a certain region of space, we can assign to every point the value of a physical quantity (such as temperature), we call that particular arrangement or distribution of the quantity a field. For instance, we may have a temperature field, where the temperatures in a given plane can be represented on paper by isothermal lines which are drawn through all points that have the same chosen temperature. We may have a pressure field, similarly represented by isobars (Figure 4.1a). The two examples mentioned are called *scalar fields* since temperature and

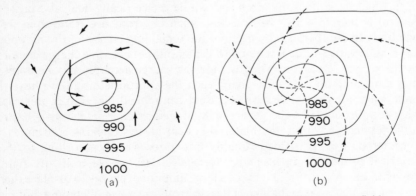

Figure 4.1. Examples of a scalar field (air pressure) and vector field (wind velocity). Full lines are isobars. Wind velocity is represented (a) by vectors and (b) by stream lines (dotted)

pressure are scalar quantities (Chapter 1), but it is also possible to have *vector fields*, where the quantity specified at every point is a vector (such as wind velocity). Vector fields are represented on paper either by small vectors drawn at selected points in the field (Figure 4.1a) or by *stream lines* whose direction at any point is that of the vector at the same point of the field (Figure 4.1b).

4.1 Stationary Electric Charges

When a comb is run through dry hair the hair is attracted to the comb, and when taking off a nylon garment it is often found to cling to the body. The separated objects have been *charged* by the effects of friction, and one object is said to have acquired an electric charge of opposite sign to that acquired by the other, one category of charge being called positive and the other negative. Charges of the same sign repel each other, whereas charges of opposite sign are attracted together. Electric charges therefore exert forces upon each other, even at a distance, in the same manner that gravity acts between bodies that are not in contact (with the difference that gravitational forces are always attractive). The amount of charge acquired by an object can be measured, as we shall see, from the force it exerts on another charge in given circumstances. In experiments on charging by friction, Faraday showed that in any such process equal amounts of positive and negative charge were always produced and that when equal charges of opposite sign were brought together the result was a disappearance of any charge phenomena.

The modern theory of the nature of matter explains Faraday's results, for we consider an atom to consist of a positively charged nucleus surrounded by negative electric charges—electrons. The total charge of the atomic electrons is generally equal in magnitude, though opposite in sign, to that of the nucleus, so that the atom as a whole is electrically neutral. Matter in general is therefore also neutral, but when electrons are transferred from one body to another (e.g. by friction) the remaining charge of the first body has a net positive value just equal to the amount of negative charge acquired by the second. Note that when we add the charge acquired by one body to that remaining on the other, giving each the appropriate sign, we have the same value for the net charge as existed beforehand (i.e. zero in this case). This is an example of the general law of *conservation of charge*—that the total net charge within any material system always remains constant.

Materials can be divided roughly into three classes according to their electrical properties. In *insulators* (or *dielectrics*) the electrons are tightly bound in the structure and it takes a large amount of energy to displace them. In *semiconductors* (see Chapter 6) the electrons are normally bound, but may be displaced by relatively small amounts of energy to energy levels at which they are no longer tied to a particular atom and can move through the material. *Conductors*, such as metals, have, on the other hand, a permanent number of electrons which are free to move at random among the atoms; these constitute an electric current if a drift in one direction is superimposed on their random motions. Current is defined as the rate of flow of charge per second.

4.1.1 *The Force between Charges*

Suppose the force exerted on a charged body, due to the presence of other charges, is twice as great as that exerted on another similar body placed in exactly the same situation. We then say, by definition, that the charge on the first body is twice that on the second (i.e. force is proportional to charge). On the other hand, if we move one charged body around in the presence of another stationary one, we find the direction and magnitude of the force exerted depend on the position of the body in space. To each point in space we can therefore assign a vector which describes the direction and magnitude of the force per unit charge that would act on a charge placed at that point. (We assume that this charge is so minute that it does not disturb the situation we are describing, i.e. it is a 'point' charge.) This force per unit charge is called the *electric field intensity*, **E**, and its distribution in space is the electric field. When we know the field intensity at a given point we can estimate the force **F** exerted there on a charge of value e from the equation

$$\mathbf{F} = e\,\mathbf{E}. \tag{4.1}$$

The equation is written so that if e is positive, the force vector **F** has the same sign (i.e. points in the same direction) as the vector **E**; thus a positive charge tends to move along the direction of **E**, but a negative one experiences a force in the opposite direction (Figure 4.2). The electric field in the neigh–

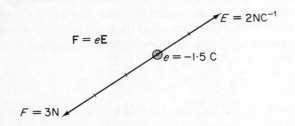

$E = 2\text{NC}^{-1}$

$\mathbf{F} = e\mathbf{E}$

$e = -1 \cdot 5$ C

$F = 3\text{N}$

Figure 4.2. The force on a charge in an electrostatic field

bourhood of two oppositely charged bodies is shown in Figure 4.3. The stream lines of an electric field are called the *lines of force*, since they show the direction of the force on a positive charge.

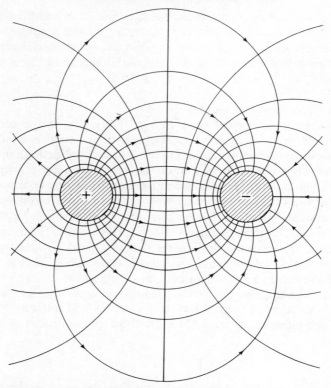

Figure 4.3.   A section of the electric field between two long charged
rods. The lines of force (arrowed) are perpendicular to the equipotentials

The nature of the force between two very small charged bodies was
investigated by Coulomb, and his findings, extended to an idealized situation
in which the bodies are infinitely small, are summarized in Coulomb's law,
or the inverse square law. This states that the force between two such 'point'
charges of magnitudes e_1 and e_2, separated by a distance r, is given by

$$F = \frac{e_1 e_2}{k r^2},\qquad(4.2)$$

where k is a constant and the direction of **F** is along the line joining the

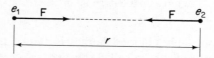

Figure 4.4.   The force between two point charges of opposite signs

two charges (Figure 4.4). The force on each charge thus falls off inversely with the square of their separation. If the size of the unit of charge has already been decided upon, so that e_1 and e_2 can be measured, then in principle an experiment can be done to find the value of k in equation (4.2), since F and r can also be measured. Usually k is determined for the case where the two charges are in a vacuum (i.e. in 'free space').

The S.I. unit of electric current, the ampere (A), is the basic electric unit, while the unit of charge, the coulomb (C), is the charge transported in one second by a current of one ampere. Since we measure F in newtons and r in metres, then k must be expressed in $C^2 m^{-2} N^{-1}$ to fit the form of equation (4.2). In practice, it is convenient to replace k by the expression $4\pi \varepsilon_0$, where $\varepsilon_0 = k/4\pi$. (This has the effect of simplifying other equations of electrostatics.) We then have

$$F = \frac{e_1 e_2}{4\pi \varepsilon_0 r^2} \quad in \ vacuo \qquad (4.3)$$

in which ε_0 is called the permittivity of free space and has the value $8\cdot854 \times 10^{-12} C^2 m^{-2} N^{-1}$.

4.1.2 Field and Potential

If we compare equations (4.1) and (4.3) we see that there are two ways of expressing the force on one point charge due to another—either explicitly (4.3) or implicitly, where we assume the charge produces a field which then exerts a force on the other charge (4.1). Since both equations must give the same expression for F, the field E produced at a distance r by a point charge e in vacuo must be given by

$$E = \frac{e}{4\pi \varepsilon_0 r^2}. \qquad (4.4)$$

The direction of the vector $\mathbf{E}$ is radially outwards from a positive charge, since one positive charge repels another positive one.

If we had a positive charge in an electric field, and allowed the charge to move freely, it would accelerate in the direction of $\mathbf{E}$ in the same way that a dropped weight accelerates in the direction of a gravitational force. In both cases the object considered has an initial *potential energy*, since the force on it is capable of doing physical work if we allow the object to move. As the object moves from one position to another, the work done by the force is defined to be equal to the decrease in potential energy. In the gravitational case, the potential energy of an object is taken to be zero at an arbitrarily chosen position which we call ground level; the potential energy of the object at another position P is the work that would be done by gravity if the object moved from P to ground level (Figure 4.5).

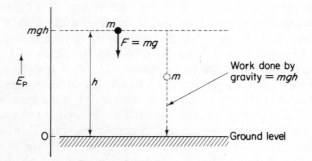

Figure 4.5.   Gravitational potential energy of a mass m at a height h above ground level

In the electrical case, the potential energy of a charge is said to be zero when it is at an infinite distance from all other charges; the potential energy at P is the work that would be done by the electric field if the charge moved away from P to such an infinite position. Since the force on a charge is proportional to the value of the charge, the potential energy of a charge depends on its value as well as its position. However, we can find for each point in an electric field a *potential energy per unit charge*, in joules per coulomb. This value is usually just called the *potential* of the point in question, and has a named unit, the volt (V). (Thus, $1V = 1J\ C^{-1}$.) The potential energy of a charge e at a point where the potential has a value ϕ is therefore given by

$$E_p = e\phi, \tag{4.5}$$

where E_p is in joules if e is in coulombs and ϕ in volts.

As an example, let us find the potential ϕ_A at a point A distant r_A from a point charge e *in vacuo*. We consider a charge q at A and calculate the work the field would do on it if it moved away an infinite distance to a position of zero potential. The force per unit charge (E) at a distance r is given by equation (4.4), and the force on q is Eq. If q increases its distance r by a small amount dr, the work done (force × distance) is given by

$$dW = (Eq)dr = \frac{eq}{4\pi\ \varepsilon_0}\frac{dr}{r^2}.$$

The total work W is found by integrating dW as the charge q moves from $r = r_A$ to $r = \infty$:

$$W = \int_{r_A}^{\infty}\frac{eq}{4\pi\ \varepsilon_0}\frac{dr}{r^2} = \frac{eq}{4\pi\ \varepsilon_0}\left(\frac{1}{r_A}-\frac{1}{\infty}\right),$$

that is

$$W = \frac{eq}{4\pi \, \varepsilon_0 \, r_A}$$

and $\quad \phi_A = \dfrac{e}{4\pi \, \varepsilon_0 \, r_A} \qquad$ *in vacuo,* $\qquad\qquad$ (4.6)

since ϕ is a potential energy per unit charge.

Note that ϕ is a scalar quantity, and so we can simply add together at a point the values of ϕ produced there by the presence of different point charges at different distances, to obtain the total potential. Since any arrangement of charge can be thought of as a number of point charges in different positions, equation (4.6) is, in principle, sufficient to calculate the potential at any point in space due to any charge distribution.

The distribution of potential values in space is a scalar field, and can be represented by lines or surfaces, called equipotentials, which pass through all points at a given potential. From equation (4.6) we see that the equipotential surfaces due to a point charge are spheres (Figure 4.6), since all points in space at the same radius from the charge (i.e. on a sphere) have the same potential.

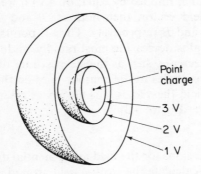

Figure 4.6. Equipotentials describing the field of a point charge. Only half of each spherical surface is shown

We now have two ways of describing the electric field in a given region of space—we can specify the value of **E** at each point, or the value of ϕ, and the two quantities are interrelated. Suppose we have a unit positive charge which moves a small distance ds from a point where the potential is ϕ to one where the potential is $\phi + \mathrm{d}\phi$ (Figure 4.7), and suppose that the change in the electric field vector **E** is negligible over the distance ds. The work done on the unit charge by electric forces can be found in two ways: (a) by determining the *decrease* in potential over the distance ds and (b) by

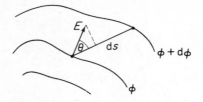

Figure 4.7.   The relation between field intensity and potential

calculating the effective force on the unit charge in the direction of movement, multiplied by the distance it moves. Thus

$$\text{(a) } dW = \phi - (\phi + d\phi) = -d\phi$$
$$\text{(b) } dW = E \cos \theta \, ds$$

$$\therefore E \cos \theta = -\frac{d\phi}{ds}. \tag{4.7}$$

In other words, the component of the field **E** taken along any direction is equal to the rate of *decrease* of potential with distance in the same direction. (Notice that we can therefore express E in volts per metre, as well as in newtons per coulomb as implied by equation 4.1.) If we consider a direction along the electric field vector, then both $\cos \theta$ and $-(d\phi/ds)$ have their maximum values, 1 and E respectively. Thus **E** points along the direction in which the potential is decreasing most rapidly, and the size of the vector equals this rate of decrease. Just as the steepest rate of descent on a hill is in a direction at right angles to the contour lines on the map, so in a 'map' of the electrostatic field the vector **E** is always at right angles to the equi–potential surfaces (see Figure 4.3).

4.1.3 *Dipoles*

By using the expressions for the field and potential due to a point charge we can, in principle, calculate the electric field around any known arrangement of charges. However, one arrangement occurs so often that it is convenient to know the field it produces, instead of calculating it when necessary. This arrangement is where two 'point' charges are separated by a distance so short as to be negligible, compared with the distance from them at which we wish to know the field.

Such a *dipole* occurs, for instance, in the hydrogen chloride molecule, where the 'centre of balance' of the positive charges of the nuclei does not quite coincide with that of the electrons (Figure 4.8). The molecule acts like two equal and opposite charges, fairly rigidly separated by a small distance. A molecule that permanently acts as a dipole in this way is called a *polar* molecule, in contrast to a non–polar one in which the positive and negative charge centres coincide unless they are separated by an electric field. A metal rod used as a radio aerial acts like a dipole when it has an

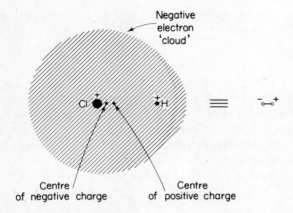

Figure 4.8.   Schematic diagram of the hydrogen chloride molecule and its dipole effect

excess of electrons at one end of the rod and a shortage at the other, so that opposite ends are oppositely charged.

The potential due to a dipole can be found from Figure 4.9. At the point

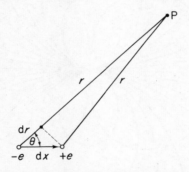

Figure 4.9.   The potential due to a dipole

P, the potential ϕ due to the charges $-e$ and $+e$ is given, from equation (4.6), by

$$\phi = \frac{1}{4\pi \, \varepsilon_0}\left[\frac{-e}{r+dr}+\frac{e}{r}\right]$$

$$= \frac{1}{4\pi \, \varepsilon_0}\frac{e\,dr}{(r+dr)r}$$

$$\simeq \frac{e \, dr}{4\pi \, \varepsilon_0 \, r^2} \, ,$$

if $dr \ll r$. Now dr is given by $dx \cos \theta$, and we call $e \, dx$ the *dipole moment* (μ) of the dipole. Thus

$$\phi = \frac{\mu \cos \theta}{4\pi \, \varepsilon_0 \, r^2} \, . \tag{4.8}$$

The dipole moment (charge × distance) is given in C m or, for molecules, which have very small dipole moments, in Debye units where 1 Debye = $3 \cdot 33 \times 10^{-30}$ C m. Notice that the potential depends on the angle between the line to P and the dipole direction (from negative to positive) and that it decreases more quickly with distance than the potential due to a simple charge—$1/r^2$ compared with $1/r$. The field of a dipole is shown in Figure 4.10.

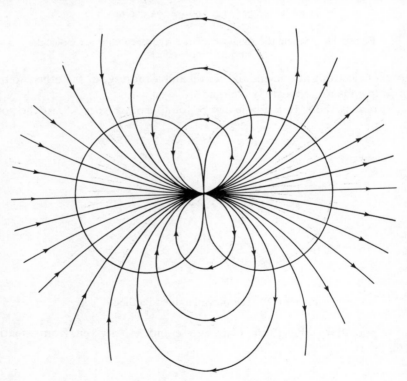

Figure 4.10. The field produced by a dipole of negligible dimensions. The dipole is at the centre of the diagram and its direction (negative charge→positive charge) points norizontally to the right. Field lines are arrowed; only one equipotential is drawn, for clarity

When a dipole is situated in what we call a uniform field (i.e. one where the lines of force are all parallel and the equipotentials are plane surfaces), the charges of the dipole experience forces which can easily be found (Figure 4.11). A force eE acts on the positive charge and a force of $-eE$ on the negative one. The two forces, acting in exactly opposite directions but,

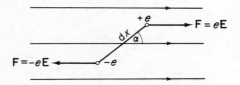

Figure 4.11.   The forces on a dipole in a uniform field

in general, along different lines of action, tend to turn the dipole and produce a couple, or torque (Chapter 1), of value $eE(\mathrm{d}x \sin \alpha)$. Substituting $e\,\mathrm{d}x = \mu$, the couple T is given by

$$T = \mu E \sin \alpha \qquad (4.9)$$

and is only reduced to zero when the dipole has turned to point along field lines. Polar molecules in a solution can be lined up in this way by applying an electric field across the liquid.

4.1.4 Charges in Conductors

When we pour some water into a glass the water molecules arrange themselves in such a way that they are in equilibrium and the gravitational forces at any part of the liquid are balanced by the liquid pressure. In the same way, when we place a conductor in an electric field the electrons move until they are in equilibrium and the force on any electron due to the new arrangement of charges just balances the effect of the applied electric field. The net force on any charge in the conductor is then zero; in other words, the *field intensity inside a conductor is zero* under static conditions. Since the field intensity is also the rate of decrease of potential with distance, a zero value implies that there is no change of potential anywhere in the conductor. The whole of the conductor has the same potential, and in particular *the surface of a conductor is an equipotential surface*.

The situation often arises in which a number of conductors at different potentials are placed close to each other. A simple example is the parallel arrangement of two flat plates in an electrostatic capacitor. Since charges are free to move in a conductor, we can rarely specify, at the outset, the distribution of charge on the various conductors. Coulomb's law cannot therefore be used directly to calculate the field due to the conductors, except in trivial cases. A more convenient expression of the inverse square law in this situation is Gauss's law, which describes the electric flux due to a collection of charges of known magnitude but of unknown distribution. The *electric flux* through a small area dA of a surface is defined to be the area multiplied by the component of **E** taken along a line perpendicular

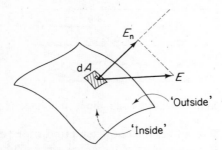

Figure 4.12.   The flux of **E** through a small area of surface. d$N = E_n$ dA

to the surface (Figure 4.12). This component is called the normal component (E_n) of **E** so that the quantity of flux (dN) through a small area is given by

$$dN = E_n \, dA. \tag{4.10}$$

The value of dN is taken as positive if **E** points from what we choose as the 'inside' of the surface to what we choose as the 'outside'.

The total flux out of a spherical surface drawn in space with a point charge at its centre is easily found. The charge e gives a field intensity $E = e/(4\pi \, \varepsilon_0 \, r^2)$ at the surface of a sphere of radius r. This field is directed outwards normal to the surface, so that $E_n = E$. Everywhere on the surface E_n has the same value, so the total flux N out of the sphere is E_n multiplied by the total surface area, $4\pi \, r^2$. Thus

$$N = \frac{e}{4\pi \, \varepsilon_0 \, r^2} \cdot 4\pi \, r^2,$$

that is

$$N = \frac{e}{\varepsilon_0}. \tag{4.11}$$

The total flux thus depends only on the amount of charge. If we consider a volume with a more complicated shape than a sphere, we can always draw round the charge a sphere which is entirely inside the volume and another one which entirely surrounds it (Figure 4.13). The flux through both spheres is e/ε_0, so it is plausible (and can be proved) that the total flux through the closed surface between them has the same value. The

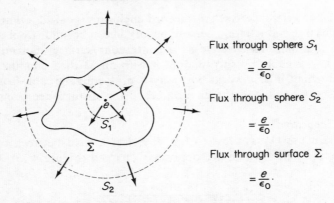

Flux through sphere S_1

$$= \frac{e}{\epsilon_0}.$$

Flux through sphere S_2

$$= \frac{e}{\epsilon_0}.$$

Flux through surface Σ

$$= \frac{e}{\epsilon_0}.$$

Figure 4.13.   The flux through any closed surface Σ due to a point charge in a vacuum

flux through *any* closed surface is thus given by equation (4.11) and if the surface encloses several point charges, each will produce its own flux; if the total charge enclosed is q, then the total flux is

$$N = \frac{q}{\varepsilon_0} \quad in\ vacuo. \tag{4.12}$$

This is a statement of Gauss's law.

One of the useful properties of a closed conducting surface is that it can act as an *electrostatic screen* (Figure 4.14). If we apply Gauss's law to a surface drawn as shown within the material of the conducting box, we

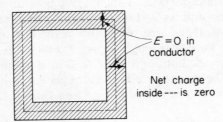

$E = 0$ in conductor

Net charge inside --- is zero

Figure 4.14.   Closed conducting surface as an electrostatic screen. The diagram shows a section through a metal box

know the field at the surface (i.e. in a conductor) is everywhere zero and therefore the flux through this surface is zero. From equation (4.12) this implies there is no net charge within the surface. Thus, for example, if there are no separated charges (and no fields) inside the box, the application of changing electric fields or voltages outside the box will not affect the situation inside, and electrical apparatus inside the box is 'screened' from outside electrical interference. Such a screened enclosure is called a 'Faraday cage', and one large enough for a person to work in can be made from wire netting (which is almost as effective as a continuous surface). Screening can be applied to a cable by surrounding it with an earthed conducting sheath. The screening, in fact, works both ways—protecting the cable from 'pick–up' of outside signals and the environment from disturbance from the cable.

4.1.5 *Capacitors*

We can use Gauss's law to find the electric field between two parallel conducting plates which are oppositely charged (Figure 4.15). Away from

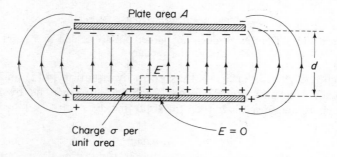

Figure 4.15.   Section through a parallel–plate capacitor carrying a charge q at a potential difference V

the edges of the plates the equipotential surfaces are planes, with the lines of force perpendicular to them. Practically all the charge of each plate is attracted to the inner surface, as shown. We now draw a box with its top between the plates and its bottom inside one conductor, and find the flux that issues from it. If the area of the top of the box is S, the flux through it is ES. There is no flux through the sides, which are parallel to the field, and none through the bottom, since $E = 0$ in a conductor. ES is the total outward flux. If the charge per unit area of plate is σ in the neighbourhood of our 'box', the charge in the box is σS and Gauss's law gives

$$ES = \frac{\sigma S}{\varepsilon_0}$$

or

$$E = \frac{\sigma}{\varepsilon_0}, \qquad (4.13)$$

which is the field between the plates.

In a system where the plates are very close together, the percentage area disturbed by 'edge effects' is small, and we can then assume $\sigma = q/A$ where q is the charge on one plate and A its area. If the separation of the plates is d, the difference in potential (V) between them is Ed, since E gives the potential change per unit distance. We thus have

$$V = Ed = \frac{q/A}{\varepsilon_0} \cdot d$$

or

$$V = q/C, \tag{4.14}$$

where C is called the *capacitance* of the system. (In this case of the parallel-plate *capacitor* we have described, $C = \varepsilon_0 A/d$ when the plates are separated by a vacuum.) Rewriting equation (4.14) in the form $q = CV$ we see that C is the charge stored on a capacitor per unit potential difference, and can be measured in coulombs per volt. In fact, the capacitance C has a special unit, the farad (F), where $1\text{F} = 1\text{C V}^{-1}$. Most capacitors used in electrical devices have capacitances with values which range from picofarads ($1\text{ pF} = 10^{-12}\text{ F}$) to thousands of microfarads ($1\text{ }\mu\text{F} = 10^{-6}\text{ F}$).

Variable capacitors are made in the form of the parallel-plate system we described, with arrangements for varying either the area of overlap (A) between the plates or the separation (d) between them. Large fixed capacitances can be produced by having a very thin ribbon of dielectric (Section 4.1.6) coated on each side with a metal film or foil; the whole is rolled up, with more insulation, into a cylinder and terminals are connected to the two films. Any two separated conductors which can be oppositely charged will act as a capacitor, and often such capacitances are unwelcome in the electrical system. For instance, the microelectrodes used in measuring nerve potentials are finely tapered glass capillaries filled with a conducting electrolyte which links the nerve to the measuring apparatus. The electrolyte, separated from the electrolytes of the tissue by a thin layer of glass, forms an unwanted capacitor with a capacitance of the order of 5 pF.

The effect of linked capacitors in a circuit depends on whether they are connected in parallel or in series. In parallel (Figure 4.16a) the capacitors

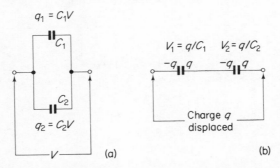

Figure 4.16. Combination of capacitors (a) in parallel and (b) in series

have a common potential difference V but different charges, $C_1 V$ and $C_2 V$. The equivalent capacitance C' is given by total charge/potential difference:

$$C' = \frac{C_1 V + C_2 V}{V} = C_1 + C_2. \tag{4.15}$$

In series (Figure 4.16b) the charge q displaced from one capacitor to the next is the same for all capacitors, but they have various potential differences, q/C_1 and q/C_2. Then (since only a charge q has been displaced),

$$\frac{1}{C'} = \frac{q/C_1 + q/C_2}{q} = \frac{1}{C_1} + \frac{1}{C_2}. \tag{4.16}$$

4.1.6 Dielectrics

If we put an insulator or *dielectric* in an electric field (e.g. between the plates of a charged capacitor), the field distorts the atoms and molecules so that their positive and negative charge centres are separated and they act as small dipoles. The permanent dipoles of polar molecules, initially pointing in random directions, tend to orient themselves along the field lines. These two effects, called *distortion polarization* and *orientation polarization*, affect the original electric field in the material. Faraday found that completely filling the space between the electrodes of a charged capacitor with a dielectric instead of vacuum reduced the field (and hence the potential difference) by a factor ε_r, which he called the dielectric constant of the material. (Note that the capacitance $C = q/V$ is *increased* by the factor ε_r, since V is reduced for a given value of q.) Another name for ε_r is the *relative permittivity*; values are given in Table 4.1.

TABLE 4.1
Values of the permittivity of common substances

Substance	Relative permittivity or Dielectric constant ε_r	Absolute permittivity $(C^2\ m^{-2}\ N^{-1})$ $\varepsilon \times 10^{12}$
Vacuum	1	8·854
Air	1·00059	8·854
Benzene (0°C)	2·3	20·4
Mica	6·0	53
Ethanol (0°C)	28·4	251
Water (20°C)	80	709

It follows that the electric field calculated for a vacuum (e.g. between the plates of a parallel–plate capacitor—equation 4.13) will be reduced in a dielectric:

$$E = \frac{1}{\varepsilon_r} \left(\frac{\sigma}{\varepsilon_0} \right)$$

or

$$\varepsilon_r \, \varepsilon_0 \, E = \sigma \tag{4.17}$$

for a dielectric of relative permittivity ε_r. The product $\varepsilon = \varepsilon_r \, \varepsilon_0$ is called the absolute permittivity of a material, and εE is called the *displacement* (**D**) in the material.

Since the reduction of electric field is due to the presence of dipoles lying along the field lines, a material containing polar molecules with permanent dipole moments has a greater effect than a non–polar one, and thus a greater value of ε_r. However, the polar molecules tend to be disturbed from their orientation in the field by random molecular collisions and vibrations which have increasing effect as the temperature rises. The permittivity of a polar substance thus decreases with temperature increase.

The extra permittivity of the polar material also decreases if a continually reversing field is applied instead of a constant one. As the frequency of the alternating field is increased, the permanent dipoles have less and less time to turn round to point in the changing field directions. When the period of oscillation of the field nears the *relaxation time* of the molecules (a measure of their minimum 'turn–round' time), they can no longer follow the oscillation and the permittivity due to orientation polarization vanishes. As an example, for water in a direct field $\varepsilon_r = 80$ but in an alternating field of frequency above 16·8 GHz it falls to a value of about 2.

4.2 Charges Moving at Constant Speed

Until now we have only considered the forces on a stationary charge, but a moving charge is subject to additional forces due specifically to its motion. For example, if we bring a magnet near a television tube, the beam

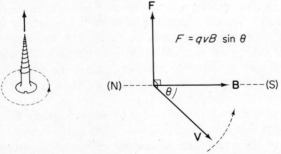

Figure 4.17.   The force on a charge moving in a magnetic field

of moving electrons is deflected and the picture is distorted. The incidence of cosmic rays is greater on the earth at higher latitudes, because the charged particles are deflected by the earth's magnetic field.

It is found that a positive charge q moving with a speed v at an angle θ to the N→S stream lines of magnetic field (Figure 4.17) experiences a sideways force, F, as shown. The force vector is perpendicular to the plane in which the field and velocity vectors lie, and points in the direction a right-hand screw would go if we turned it the same way we need to turn v, with minimum rotation, to point along the field. The magnitude of the force is given by

$$F = q\,v\,B \sin \theta, \tag{4.18}$$

in which B is the strength of the magnetic field and is called the *magnetic induction*, or *magnetic flux density*. If F is in newtons, q in coulombs and v in metres per second, the unit of B is $1\ \mathrm{N\,s\,C^{-1}\,m^{-1}}$ and in the S.I. system is given the name tesla (T). (This unit is equivalent, as we shall see later, to the unit of 1 weber per square metre ($1\ \mathrm{Wb\,m^{-2}}$) used in the M.K.S. system.) When the charges are moving in a conductor (e.g. a wire) of length l the charge q in that particular length at a given instant moves out of it in a time $t = l/v$, so that the current $I = q/t = qv/l$. We can thus replace qv in equation (4.18) and obtain

$$F = I\,l\,B \sin \theta \tag{4.19}$$

for the force on a length of conductor at an angle θ to a uniform magnetic field. Measuring the force on a conductor in a constant magnetic field is a good way of finding the current in it, and this is the basis of the moving–coil (m.c.) meter which finds uses in all kinds of electrical equipment.

In the moving–coil meter (Figure 4.18) a coil of fine wire is wound on a rectangular frame pivoted at its ends, so that the coil can rotate in the field produced by a permanent magnet. Current is led through the coil via hair-springs attached to the body of the instrument, which also serve to restrain the rotation. The field is designed to be radial, so that current flowing along the two sides of the rectangle is always at right angles to the field and $\sin \theta = 1$. Current flowing along the ends of the rectangle is parallel to the field ($\sin \theta = 0$) and experiences no force. The forces on the sides of the rectangular coil produce a couple, proportional to the current, which rotates the coil and its attached pointer against the springs. Very finely balanced systems can be made which take as little current as 50 μA to produce a full–scale deflection (F.S.D.).

Besides being subject to the forces due to other magnetic fields, a moving charge or a current produces a magnetic field itself. In fact, all magnetic fields are due to the circulation of charges, either on a macroscopic scale (e.g. around coils of wire) or on a microscopic one (e.g. around orbitals

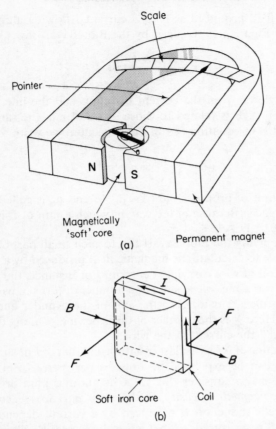

(a)

(b)

Figure 4.18. (a) Construction of the moving–coil meter (simplified). (b) The couple on the rectangular coil, due to the current I in the field B

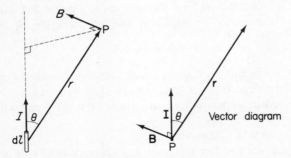

Figure 4.19. The magnetic field produced by a current flowing in a length dl of a conductor

of an atom). The field produced at P by a current I flowing along a short path of length dl (Figure 4.19) is given by the Biot–Savart law,

$$B \propto \frac{I \, dl \, \sin \theta}{r^2} \, ,$$

where θ is the angle between the current direction and the line $\mathbf{r}$ from dl to P. The direction of $\mathbf{B}$ is related to those of $\mathbf{I}$ and $\mathbf{r}$ in the same way $\mathbf{F}$ is related to $\mathbf{v}$ and $\mathbf{B}$ in equation (4.18). For a vacuum we write

$$B = \frac{\mu_0}{4\pi} \left(\frac{I \, dl \, \sin \theta}{r^2} \right), \qquad (4.20)$$

in which the constant of proportionality is $\mu_0/4\pi$ and μ_0 is called the permeability of free space. (Because of the way in which a unit of B is defined, $\mu_0/4\pi$ is exactly 10^{-7} N A^{-2}.)

Obviously by applying the Biot–Savart law to each small part of a circuit in turn, it is possible to calculate the magnetic field produced by a complete circuit, such as a coil of wire carrying a current. For instance, the magnetic induction inside a very long *solenoid* (a large number of turns of wire wound on a cylindrical former) is parallel to the axis of the cylinder and has the value $B = \mu_0 \, NI$. Here N is the number of turns per metre along the length of the solenoid and I the current in the wire.

Notice that the value of NI in the last equation and that of $I \, dl \, \sin \theta/(4\pi \, r^2)$ in equation (4.20) can be expressed in amperes per metre (A m^{-1}); it is sometimes convenient to consider this value as another kind of magnetic field (H) called the *magnetic intensity*, so that $B = \mu_0 H$ in a vacuum. The value of the magnetic induction B produced by a current depends on the material in which B is measured, so that generally $B = \mu \, H$, where μ is the magnetic permeability of the material. In ferromagnetic materials such as iron and magnetic alloys, B is substantially increased and the relative permeability μ_r (i.e. μ/μ_0) can have values as great as 10^5.

Just as in electrostatics we have the concept of electric flux (equation 4.10), so in the magnetic field the *magnetic flux of induction*, Φ, is defined by

$$d\Phi = B_n \, dA, \qquad (4.21)$$

in which dΦ is the flux through a small area dA due to the normal component B_n of the field $\mathbf{B}$. Flux Φ is measured in webers (Wb) and hence B can be expressed in webers per square metre (Wb m^{-2}) as well as in teslas.

4.3 Steady Currents in Conductors

Normally, an electron free to move in an electric field will be accelerated by the force acting on it, and its velocity will not be constant. However, if electrons are moving in a medium (such as a metal conductor) where they make collisions due to random thermal motions, their forward velocity acquired by accelerating after one collision tends to be destroyed at the next, so that their *average* drift velocity in the direction of the field is constant, and we have a steady direct current (D.C.) if the field and supply of electrons is constant. (The existence of a field in the conductor does not contradict our earlier statement that $E = 0$ under conditions of *static* charge.) The current is measured in coulombs per second or amperes (A), and is taken to flow in the direction of movement of positive charges, although most current (except, for example, for positive ions in an electrolyte) is carried by negative electrons moving in the opposite direction to the conventional current.

For many conducting materials the current flow per unit area (I/S) is proportional to the applied electric field, whatever its value, and the constant (σ) of proportionality, called the *conductivity*, is independent of the field, although it may depend on temperature. We may therefore write

$$\sigma = \frac{I/S}{E}. \tag{4.22}$$

Suppose we have a conductor of constant cross–sectional area S and length l with a potential difference V between its ends. The field E in the conductor is V/l volts per metre, so from equation (4.22) we obtain

$$V = \left(I\frac{l}{\sigma S} \right) = IR, \tag{4.23}$$

where R is the *resistance* of the conductor, measured in ohms (Ω). If σ is

Black	0	Green	5
Brown	1	Blue	6
Red	2	Violet	7
Orange	3	Grey	8
Yellow	4	White	9
Gold	±5%	Silver	±10%

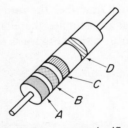

Figure 4.20. The resistor colour code. A denotes most significant figure of resistance in ohms. B denotes second significant figure of resistance in ohms. C denotes number of zeros to follow figures A and B. D (if present) denotes the tolerance by which the resistance may deviate from the marked value. No band indicates a tolerance of 20%. E.g. if A,B,C,D are respectively yellow, violet, orange and silver, then $R = 47\,000\,\Omega$ ($\pm 10\%$).

(*Note* : standard combinations of A and B increase R in steps of c. 50%, viz: 10, 15, 22, 33, 47, 68.)

independent of E, R is independent of V, for a given temperature, and materials in which this holds are said to obey Ohm's law. Generally, metals are so–called 'ohmic' conductors and obey the law, while semiconducting materials are 'non–ohmic' and do not. An element in an electric circuit which is designed to exhibit resistance and no other electrical property is called a *resistor*. For ease of recognition (e.g. in electronic circuits) resistors are usually 'colour–coded' with painted markings which denote both the nominal resistance and the tolerance by which the resistance is likely to deviate from the marked value (Figure 4.20).

The work done by a field in moving a charge dQ from one point to another through a potential difference V is $V\,dQ$. If this is done in a time dt, the work done per second (i.e. the power expended) is $V\,dQ/dt$, but since dQ/dt is the current flow I, we have:

$$\text{Power dissipation} = VI. \tag{4.24}$$

For a conductor of resistance R, we can use equation (4.23) to obtain:

$$\text{Power dissipation} = \frac{V^2}{R} = I^2R. \tag{4.25}$$

If I is in amperes and V in volts, the unit of power is one joule per second, which is given the name watt (W). Equation (4.24) gives the instantaneous rate at which electrical energy is converted into another form in any system (e.g. in an electric motor, an electrolyte, etc.) which carries a given current at a given voltage. In a conductor, the energy imparted to the electrons as they move in an electric field is given up to atoms of the material at each collision and converted to thermal energy. In this case, the power dissipation appears as heat released in the conductor ('Joule heating') and equation (4.25), which describes the rate of heat production, is known as Joule's law.

In order to estimate the effect on a circuit of several resistors, we note that when in series (Figure 4.21a) they carry a common current but when

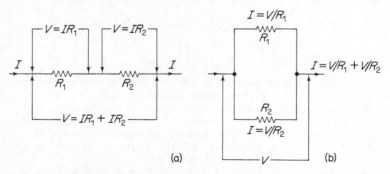

Figure 4.21. Combination of resistors (a) in series and (b) in parallel

in parallel (Figure 4.21b) they have a common potential difference. In series, the single equivalent series resistance R' is given by

$$R' = \frac{IR_1 + IR_2}{I} = R_1 + R_2. \tag{4.26}$$

In parallel,

$$\frac{1}{R'} = \frac{V/R_1 + V/R_2}{V} = \frac{1}{R_1} + \frac{1}{R_2}. \tag{4.27}$$

4.3.1 Circuit Theory

In order for current to flow continuously round a closed conducting circuit, the total power dissipated in all the various parts of the circuit must be compensated by a continuous input of electrical power from chemical, mechanical or other generators (e.g. batteries, dynamos, photocells, etc.). Since the current flows through the generator itself, we can think of it as producing an amount of power $\mathcal{E}$ times the current I that it delivers to the circuit. The instantaneous power dissipation in the circuit is the sum of all the I^2R or VI expressions for the various components of the circuit, including any power losses in the generator itself. Thus we have, for example, in a circuit where all the components are resistive and in series (Figure 4.22),

$$\mathcal{E} I = I^2 r + I^2 R_1 + I^2 R_2 + \ldots = \Sigma I^2 R$$

or

$$\mathcal{E} = \Sigma IR. \tag{4.28}$$

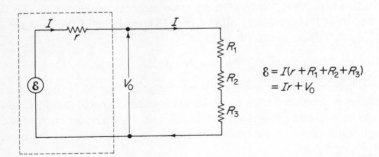

Figure 4.22.   The emf and output voltage V_0 of a generator with internal resistance r

Since IR is a potential difference, $\mathcal{E}$ must have the dimensions of potential; it is called the electromotive force (emf) acting in the circuit, and is expressed in volts.

Equation (4.28) can be extended to a complicated circuit containing a

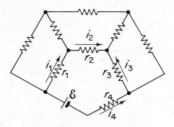

Figure 4.23.   A circuit network. Considering the loop which contains
the battery, $\varepsilon = i_1 r_1 + i_2 r_2 - i_3 r_3 - i_4 r_4$

network of conductors and several sources of emf (Figure 4.23), and the resulting law (Kirchhoff's second law) states that:

> In any closed loop of a conducting network, the algebraic sum of the emfs is equal to the algebraic sum of the IR products.

Kirchhoff's first law for a network expresses the fact that, in the steady state, charge does not accumulate at any point in the circuit. In particular, where several conductors meet (a 'node'):

> The algebraic sum of the currents flowing towards a node is zero.

The application of Kirchhoff's two laws enables us to calculate the currents in a network, given the emfs and resistances.

Notice that the emf ε is not, in general, the same as the voltage V_0 measured across the terminals of a generator, since if the generator has internal power losses, it has an effective internal resistance r. The voltage V_0 (Figure 4.22) equals the sum of the (IR) products in the *external* circuit only, so that from equation (4.28)

$$\varepsilon = Ir + V_0 . \tag{4.29}$$

Only when no current is drawn from the generator ($I = 0$) are ε and V_0 the same. A generator with low internal resistance ($r \to 0$) is called a *constant voltage* source, since Ir has very little effect on V_0 whatever current is drawn. A high value of r, on the other hand, gives a *constant current* source, since $Ir \gg V_0$ and $I \simeq \varepsilon / r$.

It can be shown that a generator or source of emf delivers the maximum power to an external circuit when the equivalent resistance of this circuit is exactly equal to the internal resistance of the generator (i.e. $R' = r$ in Figure 4.24). (Check this by calculating $I^2 R'$ for the figure, when $R' = 4$, 5 and 6 Ω.) This is an instance of a more general property, that maximum power is transferred from one part of a circuit to the other if each part separately has the same resistance, measured between its terminals. If the two parts have this property, they are said to be *matched*. (See also *impedance matching*, Chapter 5.)

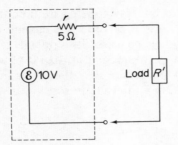

Figure 4.24. Matching the load to the generator. When R' is connected across the terminals, the maximum power is developed in the load when $R' = r$. (In this case, the maximum power is 5 W.)

4.3.2 Circuit Measurements

In building or checking circuits we frequently need to measure values of current, voltage and resistance. All three measurements can be made easily (though not always accurately) with the use of a sensitive moving–coil meter. Suppose we have such a meter which gives a full–scale deflection when taking 100 μA of current, and its coil has a resistance of 100 Ω. By the use of appropriate resistors it is possible to adapt the meter to measure currents much greater than 100 μA, voltage and, by including a battery in the circuit, resistance. We shall consider each application in turn.

To convert the instrument from a microammeter into a milliammeter, which measures up to, say, 10 mA instead of 100 μA, we have to arrange for a large proportion of the measured current to by-pass the meter. This is done by connecting a 'shunt' resistor across the meter terminals (Figure 4.25) so that when shunt and meter combined are passing 10 mA, the meter

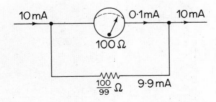

Figure 4.25. Use of a moving–coil meter as a milliammeter

takes 100 μA (i.e. 0·1 mA) and gives a full–scale deflection while the shunt takes 9·9 mA. Since we want the shunt to take 9·9/0·1 $=$ 99 times the current in the meter, its resistance must be 1/99 of the meter resistance (i.e. 100/99 $=$ 1·01 Ω). (Note that the resistance of the shunted meter (1·00 Ω) will significantly reduce the current in a circuit in which it is put, if the total resistance in the rest of the circuit is only a few ohms.)

To convert our meter into a voltmeter, we note that when it registers a full–scale deflection, a current of 100 μA is passing through the resistance of 100 Ω, and there is a voltage across its terminals of $IR = 100\ \Omega \times 100\ \mu$A $= 10$ mV. If the meter is to measure up to 1·0 V, we need to increase its effective resistance to R', where $R' \times 100\ \mu$A $= 1·0$ V (i.e. $R' = 10$ kΩ). This is done (Figure 4.26) by putting in series with the meter a resistor of

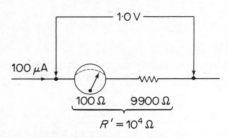

Figure 4.26.   Use of a moving–coil meter as a voltmeter

value 9900 Ω (9·9 kΩ). Although the voltmeter so constructed has a resistance of 10 kΩ, it still takes a significant current from the source of the voltage that is measured, and may thus reduce the output voltage considerably if the source has a large internal resistance, of the order of 1000 Ω or more. To make accurate measurements in such a case, other instruments than the moving–coil meter are used which take practically no current from the voltage source.

To measure resistance, the meter is connected in series with a source of emf (e.g. a dry battery) and a resistor; the circuit is completed through the resistance to be measured (Figure 4.27). It is arranged that the meter shall

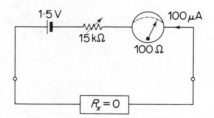

Figure 4.27.   Use of a moving–coil meter to measure resistance, R_x.
The variable resistor is adjusted so that $I = 100\ \mu$A when $R_x = 0$

give a full–scale deflection when the unknown resistance is zero, and, of course, it gives a zero deflection when the unknown resistance is infinite. The calibration of the scale in ohms therefore runs in the reverse direction to voltage and current scales, and unlike them is non–linear (i.e. equal increases in the measured resistance do not produce equal decreases in the meter deflection). Although ammeters and voltmeters are produced as separate instruments, the resistance meter is usually incorporated in the instrument called a 'multimeter', where one moving–coil meter can be internally connected by suitable switching to measure at will either amperes, volts or ohms in the way we have described.

Another device for measuring resistance, which is particularly useful in detecting small changes of resistance (e.g. in a thermistor or a resistance thermometer), is the *Wheatstone bridge*. The unknown resistor R_1 is connected into a 'bridge' network with three known resistors as shown in Figure 4.28. One of the known resistors (R_2) is variable while the other

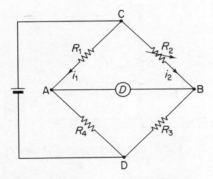

Figure 4.28.   The Wheatstone bridge

two are related by a convenient factor (e.g. $R_3 = 10\,R_4$). Two opposite corners of the network are connected to a source of emf and the other two to a voltage detector (D) such as a moving–coil meter. The value of R_2 is adjusted until points A and B on each side of the meter are at the same potential, when the detector gives a zero reading. (The Wheatstone bridge is therefore a *null–reading* device, which does not depend on the accuracy of the detector as it is only required to show zero or non–zero.) At this 'balance' position the potential drop across CA is equal to that across CB:

$$i_1 R_1 = i_2 R_2.$$

Since A and B are at the same potential no current flows in AB and all i_1 flows into R_4 and all i_2 into R_3. We then also have

$$i_1 R_4 = i_2 R_3$$

and from the two equations

$$R_1 = R_2\,(R_4/R_3). \tag{4.30}$$

When the ratio R_4/R_3 contains only powers of ten it is easy to read off the value of R_1 from that of the setting of R_2. After the bridge has been set at balance, small changes in one of the resistances produce proportional changes in the potential difference across the detector, so the bridge can be set up as a sensitive device for measuring *changes* in resistance. The most sensitive arrangement is when all four resistors and the detector have approximately equal resistances.

An important null method for measuring potential difference is the use of a *potentiometer*. It has the advantage that when the measurement is made, no current is taken from the measured source of voltage. In Figure 4.29 a

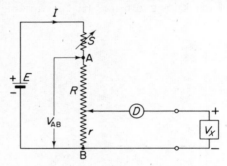

Figure 4.29.   Use of a potentiometer. *D* registers zero if
$$V_x = (r/R)V_{AB}$$

source of emf, *E*, is connected through a variable resistor *S* to a resistor *R* (such as a length of bare wire) from which a chosen resistance *r* can be selected by means of a sliding connection. The resistor *R* (the potentiometer) acts as a 'voltage divider'—the potential difference V_{AB} between the points A and B is *IR* while that between the slider and B is *Ir*, so that a fraction r/R of V_{AB} appears across *r*. Usually *r* is proportional to the length of resistor selected by the slider, and r/R can be determined. The value of V_{AB} can be made a convenient value by the selection of *E* and adjustment of *S*.

If now an unknown source of emf, V_x, is connected (in the right direction) from B through a voltage detector (*D*) to the potentiometer slider, the emf *opposes* the voltage across *r* and *D* will register zero if *r* is adjusted to make $V_x = (r/R)V_{AB}$. The unknown V_x can thus be calculated from the value of (r/R) at this 'null point' or 'balance position', at which, it should be noted, no current flows through *D* from the measured source.

Because the potentiometer takes no current when measuring a potential difference, it is suitable for measuring the emf of a system that has a large internal resistance, where the output voltage is a function of the current drawn from the system.

4.3.3 *The Pen Recorder and Negative Feedback*

Frequently it is necessary to record variations in a given quantity (pressure, temperature, nerve action potential, etc.) as a function of time. One method is to convert the variations where necessary into variations of voltage by the use of a voltage transducer. (A *transducer* is a system that provides an output of one physical quantity corresponding to the input of a different one.) The output from the transducer is amplified when necessary and applied for measurement to a potentiometer, as we have already described.

However, to make rapid measurements and record them graphically an automatic method is necessary for finding the 'balance' position of the potentiometer slider. This is done essentially by using as the detector D in Figure 4.29 a 'differential amplifier' which detects and amplifies the *difference* between the potential of the slider and that of the input to be measured. If the slider is not at the balance position the differential amplifier gives an output (the 'error' signal) which is used to control an electric motor which drives the slider up or down the potentiometer wire (Figure 4.30).

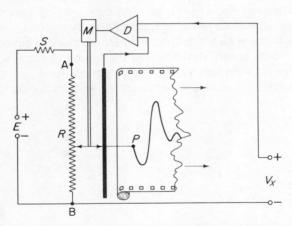

Figure 4.30.   Operation of a pen recorder (schematic). The chart pen
P is rigidly connected to the potentiometer slider, which is actuated by
motor M. The motor is controlled by the error signal from D

It is arranged that if the slider is not at the required position, the 'error signal' from D has the effect on the motor of driving the slider *towards* the right position, thus *reducing* the error signal until balance is reached. At this point the error signal is zero and the motor stops. The slider thus 'finds' the balance position automatically and the position of the slider follows any variations in the measured voltage V_x. A pen attached to the slider records its position on chart paper which is driven past the pen at constant speed, thus giving a graph of voltage versus time.

The arrangement used to maintain balance is an example of what is called 'negative feedback', in which the deviation of a system from a certain state generates an error signal, which is fed back into the system in such a sense as to reduce the deviation. (The motors which 'obey' the control signals in a mechanical feedback system are called servomotors.) A system which incorporates negative feedback is inherently stable, since any deviation from the required state is automatically reduced. A system, on the other hand, which has an amount of positive feedback is unstable, as the error signal in this case is fed back into the system with reversed sign and has the effect of increasing the deviation. This increases the error signal, which in turn causes further deviation, and the system 'runs away' from its initial state.

The principle of negative feedback is not only useful in mechanical automation. It is of great importance in electronic circuits and in any stabilized system (such as a water–bath maintained at a constant temperature). Since negative feedback is a regulatory mechanism, it is not surprising that it has a direct parallel in biological systems, where the maintenance of a stable state is called homeostasis. An organism must maintain a constant internal environment if it is to survive, so it must employ negative–feedback or homeostatic mechanisms which automatically correct any deviation. For instance, in mammals a drop in skin temperature produces a nervous 'error signal' which causes shivering and the raising of hairs, both of which tend to restore the skin temperature to its normal level.

The application of negative feedback in electronics is discussed in Chapter 6.

Further Reading

Duffin, W.J. (1965). *Electricity and Magnetism*. McGraw–Hill, New York.

Horrobin, D.F. (1970). *Principles of Biological Control*. Medical and Technical Publishing Co., Aylesbury.

Katz, B. (1966). *Nerve, Muscle and Synapse*. McGraw–Hill, New York.

Kay, R.H. (1964). *Experimental Biology*. Chapman and Hall, London.

Kip, A.F. (1962). *Fundamentals of Electricity and Magnetism*. McGraw–Hill, New York.

McFarland, D.J. (1971). *Feedback Mechanisms in Animal Behaviour*. Academic Press, New York.

Richards, J.A., F.W. Sears, M.R. Wehr and M.W. Zemansky (1960). *Modern University Physics*, Part 2. Addison–Wesley, Reading, Mass.

PROBLEMS

4.1 An ammonium ion, NH_4^+, has a net positive charge equivalent in magnitude to the charge of an electron. What is the force exerted on this ion by a triply charged PO_4^{3-} ion at a distance of 10 nm (neglecting the effects of other molecules present)? What force would be exerted on the ammonium ion by an electric field of 10^4 volts per metre? (Take the charge of an electron as -1.6×10^{-19} C.)

4.2 A polar molecule with a dipole moment of 3 Debye units lies with its axis at an angle of 30° to the lines of force in a field of 20 V cm^{-1}. Calculate the couple acting on the molecule.

4.3 A parallel–plate capacitor comprises two flat plates, 10 cm × 10 cm, separated by a dielectric of thickness 0.5 mm. What is the capacitance of the system (a) if the dielectric is air and (b) if it is mica?

4.4 A long solenoid coil is wound on a cylindrical rod of cross–section 1 cm^2 and has 5 turns per mm. If the wire carries a current of 0.5 A and the relative permeability of the rod material is 5000, calculate (a) the magnetic intensity (H) within the rod, (b) the magnetic induction, B, and (c) the magnetic flux through the rod.

4.5 A small heater coil has to be constructed from 20 cm of wire so that it operates from a 12 V battery and dissipates 5 W of power. Calculate the necessary resistance of the coil and the diameter the wire should have if its conductivity, σ, is 10^6 Ω^{-1} m^{-1}.

4.6 An audio signal generator has an output impedance of 600 Ω (i.e. the effective internal resistance is 600 Ω) and there are two pairs of headphones, one pair of resistance 3000 Ω and the other of 100 Ω. Find which pair would give the most power when connected to the generator, and the ratio of the powers developed by the two pairs.

4.7 A potentiometer system (Figure 4.29) is to be constructed with $E = 1 \cdot 40$ V and R (a dial–reading potentiometer) having a total resistance of 1000 Ω. The system is to measure potentials up to 10 mV directly, i.e. when V_x is 1 mV, the balance value of r should be 100 Ω; 2 mV should balance at 200 Ω, etc. Calculate the value of S needed in the circuit.

CHAPTER 5
Electrodynamics and Alternating Current

In the last chapter we examined the effects of stationary electric charges (electrostatics) and those due to charges moving at constant speed (e.g. the magnetic field due to a steady current). In this chapter we go on to consider the effects of *accelerating* charges, that is the effects caused by currents which are changing with time. This topic (electrodynamics) is of great importance in the production of radio waves, in the production and handling of alternating currents such as are used in the domestic mains supply, and in electronics, which is playing an increasing part in the instrumentation used in biological measurement.

5.1 Electromagnetic Induction

We have already seen that a steady magnetic field exerts a force on a moving charge (Section 4.2), but it is also found that a *changing* magnetic field will exert a force on a *stationary* charge. In other words, a changing magnetic field produces an electric field in its neighbourhood, which will induce an electric charge to move if free to do so.

A simple experiment which demonstrates the effect is shown in Figure 5.1. Circuit 1 contains a loop of wire which produces a magnetic field. The field can be varied by changing the current in the loop by means of R_1.

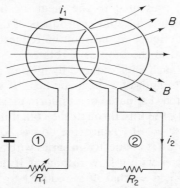

Figure 5.1. Coupled circuits which exhibit electromagnetic induction

Circuit 2 is in the magnetic field due to circuit 1 and a certain amount of the magnetic flux (equation 4.21) from circuit 1 passes through circuit 2. (The two circuits are said to be electromagnetically *coupled*.) On investigating coupled circuits of this kind, Faraday discovered that a changing current in the primary circuit (circuit 1) induced a current to flow in the secondary circuit (circuit 2). Since the current i_2 flows through the resistor R_2, an emf at least as great as $i_2 R_2$ must be acting in the secondary circuit; this is called the *induced emf* and is only present while i_1 is changing. Faraday found that the induced emf in a given circuit is proportional to the rate of change with time of the magnetic flux passing through the circuit. This statement, known as Faraday's law of induction, can be written as

$$\mathcal{E} = -\frac{d\Phi}{dt},\tag{5.1}$$

where the induced emf, $\mathcal{E}$, is in volts if the rate of change of flux, $d\Phi/dt$, is measured in webers per second.

The presence of the minus sign in the above equation indicates that the emf (and hence current) induced in a circuit is in such a direction as to produce a magnetic field which *opposes* the changes in flux which are occurring. For example in Figure 5.1, if i_1 and the magnetic flux through the secondary circuit are decreasing, the secondary current i_2 flows in the direction shown so as to produce a magnetic field which *adds* to the decreasing field from the primary circuit, and therefore tends to oppose the reduction of flux which is taking place. This principle, which follows from the law of conservation of energy, is known as Lenz's law.

Instead of changing the primary current in Figure 5.1, another way of changing the flux in the secondary circuit would be to move the circuit away to a position where the magnetic field was less. In other words, the motion of a circuit in a magnetic field may induce an emf. Yet another way of changing the flux in circuit 2 would be to rotate the loop so that the projected area it presented to the magnetic field was less. The flux through the loop, which is the sum of products of the form $B_n\, dA$ (Chapter 4) over the area of the loop, would then be reduced in magnitude and an emf would be induced in the loop as it rotated.

One application of such emfs induced by motion is in the use of a *search coil* to measure the strength of a magnetic field. If we have a small coil of wire with n turns of area A situated in a uniform magnetic field (e.g. between the poles of a magnet), the flux through the coil, if its area is perpendicular to the field lines, is $\Phi = nBA$ (since the coil is equivalent to n separate loops of area A). The coil is connected in series with a ballistic galvanometer (i.e. a moving–coil meter with a slow response). If the coil is now suddenly withdrawn from the field so that Φ is reduced to zero, a change of flux nBA occurs in the circuit before the galvanometer has time to respond.

The impulse delivered to the galvanometer by the induced flow of charge in the circuit produces a deflection proportional to the change in Φ, and therefore to B. The system is calibrated by performing the experiment with fields of known value.

5.1.1 *The Dynamo*

The continuous rotation of a coil in a magnetic field will result in the flux through the coil alternately increasing to a positive value and decreasing to a negative one, as the flux enters first from one side of the coil and then from the other. This is obviously a method of generating emfs by mechanical means (i.e. of turning work into electrical energy) and is the principle of the dynamo.

Consider for simplicity (Figure 5.2) a coil of rectangular cross–section with area A which rotates as shown at a rate ω radians per second in a uniform magnetic field B. The induced current can be led off from the rotating coil through carbon rods ('brushes') which make contact with the rotating copper slip rings S. At an instant when the normal to the coil

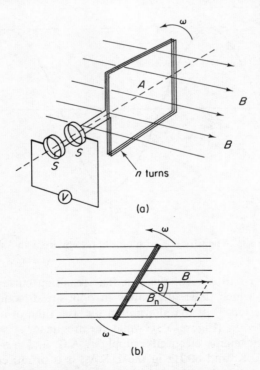

Figure 5.2.   The action of a dynamo or A.C. generator (a) in perspective and (b) viewed along the dotted axis in (a)

makes an angle θ with the magnetic field (Figure 5.2b) the magnetic field component B_n that is normal to the plane is equal to $B \cos \theta$, so that

$$\Phi = nB_n A = nAB \cos \theta \qquad (5.2)$$

is the total flux through the coil. Since $\theta = \omega t$ after t seconds, we have

$$\Phi = nAB \cos \omega t, \qquad (5.3)$$

$$\frac{d\Phi}{dt} = -\omega nAB \sin \omega t \qquad (5.4)$$

and

$$\mathcal{E} = -\frac{d\Phi}{dt} = \omega nAB \sin \omega t. \qquad (5.5)$$

The emf generated in the coil when plotted against time thus follows a sine–shaped curve (Figure 5.3) with an amplitude $\mathcal{E}_{max}$ which is proportional

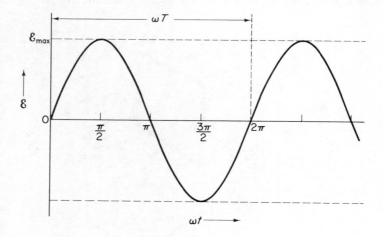

Figure 5.3.   The variation with time of the emf from an A.C. generator

to the rate of rotation ω of the coil. The resulting current is *alternating current* (A.C.), which flows alternately in opposite directions through the circuit. The period T of the alternation (i.e. the time of a whole cycle) is given by $\omega T = 2\pi$ (Figure 5.3), while the frequency $f = 1/T = \omega/(2\pi)$.

The domestic mains electricity supply is A.C. and has a frequency of 50 Hz in the U.K. and 60 Hz in the U.S.A.; it is produced by generators based on the principles discussed above. We shall return to alternating currents in Section 5.3.

5.1.2 Inductance

We saw in Figure 5.1 that a certain current in a primary circuit produced an amount of magnetic flux through a secondary circuit. In a given geo–metrical arrangement of the coupled coils, the flux through the secondary, Φ_2, is proportional to the current in the primary, i_1, so we can write

$$\Phi_2 = M_{12} i_1, \tag{5.6}$$

where M_{12} is a constant, the flux in 2 per unit current in 1. M_{12} depends on the number of turns in the coils, their shape, size, separation and the material in which they are situated. We could also write

$$\Phi_1 = M_{21} i_2 \tag{5.7}$$

for the flux in coil 1 due to a current i_2 in coil 2. We would imagine M_{21} to be another, different, constant, but, in fact, M_{12} and M_{21} have the same value, M, which is called the *mutual inductance* of the coupled coils. From equations (5.1) and (5.6) the emf, $\mathcal{E}_2$, induced in a secondary circuit is given by

$$\mathcal{E}_2 = -\frac{d\Phi_2}{dt} = -M \frac{di_1}{dt}. \tag{5.8}$$

M is thus an emf per unit rate of change of current (V A^{-1} s) and has a named unit, the henry (H). For example, if a current changing at 1 A s^{-1} in one circuit induces an emf of 1 V in another circuit, the circuits have a mutual inductance of 1 H.

Up to now we have only considered the emf induced in one coil by changing current in another. However, Faraday's and Lenz's laws also apply to a single isolated circuit. We define the *self–inductance*, L, of a circuit as the magnetic flux through a circuit due to its own current and magnetic field, so that

$$\Phi_{self} = L i \tag{5.9}$$

where L, like M, is measured in henries. Now consider a circuit (Figure 5.4) with self–inductance L to which we apply an emf V (e.g. by connecting a battery). A current i begins to flow, which creates a magnetic field and a flux $L i$ in the circuit. As the current increases from zero, the changing flux gives rise to a *self–induced* emf in the circuit, which from Lenz's law opposes the changes introduced by connecting the battery. This 'back emf' is given by

$$\mathcal{E} = -\frac{d\Phi}{dt} = -L \frac{di}{dt}. \tag{5.10}$$

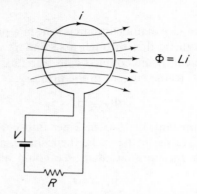

Figure 5.4.   A circuit possessing a self–inductance L

Applying Kirchhoff's second law (Chapter 4) to the circuit of total resistance R, we have

$$V - L\frac{di}{dt} = i\,R. \tag{5.11}$$

The effect of the inductance is thus to oppose the build–up of current in a circuit by reducing the effective applied voltage, and for this reason it takes a certain amount of energy to establish a current in an inductive circuit. The opposition of an inductance to change of current is analogous to the opposition of a mass to changing velocity in a mechanical system—both systems exhibit inertia.

As a simple example of the calculation of an inductance, we may take a long solenoid of small cross–sectional area A with N turns per metre and a length l. The magnetic induction in such a solenoid when carrying a current i is approximately

$$B = \mu_0\,N\,i. \tag{5.12}$$

The total magnetic flux through the turns of the solenoid due to its own field is

$$\Phi = nAB, \tag{5.13}$$

where n is the total number of turns and AB the flux through each one. From the definition of L we can therefore write

$$\Phi = Li = nA(\mu_0\,Ni),$$

but $N = n/l$, so

$$L = \mu_0 n^2\,A/l. \tag{5.14}$$

For a fixed length, the inductance of the coil is therefore proportional to the square of the number of turns.

Since L depends on flux (and hence on magnetic induction) per unit current, the inductance of a coil can be increased by filling it with a ferromagnetic material in which the value of B is substantially increased relative to that in air.

5.1.3 The Transformer

The transformer is a practical device for efficiently applying the principles illustrated by Figure 5.1, and consists essentially of two coils wound on a common former. The former usually consists of or contains a ferromagnetic material which can be easily magnetized and demagnetized without waste of energy. (Such a material is said to be magnetically 'soft'.) The purpose

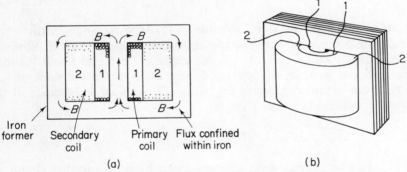

Figure 5.5.   Construction of a transformer (a) in section and (b) in perspective

of this magnetic core is to make the coupling between the coils as good as possible, that is to ensure that as much as possible of the magnetic flux produced in the core by the primary is confined to the core and passes also through the secondary coil (see Figure 5.5).

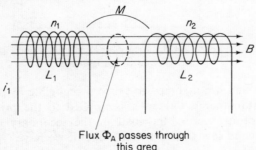

Figure 5.6.   The mutual inductance between two coils with perfect coupling. (The arc labelled M symbolizes the coupling between the coils.)

Consider two coils of a transformer, which have self–inductances L_1 and L_2 and total numbers of turns n_1 and n_2 respectively (Figure 5.6). Suppose the transformer has perfect coupling, so that the flux Φ_A which passes through the cross–section of each coil is the same, and suppose initially that this flux is due to a current i_1 in the primary coil. The total flux Φ_1 linked with the circuit of the primary coil is $n_1 \Phi_A$, since Φ_A passes through n_1 turns. The total flux is also by definition $L_1 i_1$, so that

$$\Phi_1 = n_1 \Phi_A = L_1 i_1. \tag{5.15}$$

The total flux linked with the secondary coil is similarly

$$\Phi_2 = n_2 \Phi_A,$$

that is

$$\Phi_2 = \left(\frac{n_2}{n_1} L_1\right) i_1 \tag{5.16}$$

on substituting for Φ_A from equation (5.15). The factor $(n_2/n_1)L_1$ is the mutual inductance M between the coils (see equation 5.6). If, however, we had begun with a current i_2 in the secondary coil and found the total flux produced by it in the primary circuit, we would in the same way as above have obtained a factor $(n_1/n_2)L_2$. The mutual inductance of the perfect transformer is thus given by

$$M = \frac{n_2}{n_1} L_1 = \frac{n_1}{n_2} L_2. \tag{5.17}$$

We can now predict what happens in an ideal transformer circuit, as shown in Figure 5.7, in which the coils are perfectly coupled with a mutual

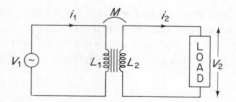

Figure 5.7. Use of a transformer. (Vertical lines indicate the coils are wound on an iron core.)

inductance M and are assumed to possess negligible resistance. A varying voltage of instantaneous value V_1 is applied to the primary coil and a voltage V_2 appears across the load in the secondary circuit. Applying Kirchhoff's second law to the primary circuit we have

$$V_1 - L_1 \frac{di_1}{dt} - M \frac{di_2}{dt} = 0, \tag{5.18}$$

where the third term is the back emf induced in the primary by changing current in the secondary coil (see equation 5.8). In the secondary circuit we obtain

$$M \frac{di_1}{dt} - L_2 \frac{di_2}{dt} = V_2, \tag{5.19}$$

where the first term is the induced emf, the second the back emf and the third the voltage across the load. In general, we would have to solve equations (5.18) and (5.19) for a given circuit. However, we will take two extreme and simple cases: (a) where the load is very small and (b) where the load is infinite (i.e. an open circuit).

Case (a). For a negligible load we can neglect V_2 in equation (5.19) which then becomes

$$\frac{di_2}{dt} = \frac{M}{L_2} \frac{di_1}{dt} = \left(\frac{n_1}{n_2}\right) \frac{di_1}{dt} \tag{5.20}$$

from equation (5.17). The system here is acting as a *current* transformer where changes in the current in the primary circuit appear in the secondary coil, multiplied by the 'turns ratio', n_1/n_2, of the two coils. A small alternating current in a primary circuit with a large number of turns will thus induce a large alternating current in a secondary circuit with few turns, connected to a small load (such as, for example, a loudspeaker).

Case (b). For an infinite load $i_2 \to 0$ and $(di_2/dt) = 0$. Equations (5.18) and (5.19) become respectively

$$V_1 = L_1 \frac{di_1}{dt},$$

$$V_2 = M \frac{di_1}{dt}.$$

Hence

$$\frac{V_2}{V_1} = \frac{M}{L_1} = \left(\frac{n_2}{n_1}\right) \tag{5.21}$$

from equation (5.17). The system now works as a *voltage transformer*, where the voltage across the large load is much greater than that applied to the primary circuit, if the secondary circuit has many more turns. This obviously can be used for voltage amplification in an electronic amplifier.

The two types of transformer are represented in Figure 5.8, together with an *isolating* transformer ($n_1 = n_2$), which is used purely to isolate a circuit (e.g. from the mains supply) so that the circuit may be earthed without disturbing the supply system. The fourth type shown is called an *auto-transformer*, in which the two circuits are not isolated and part of the primary

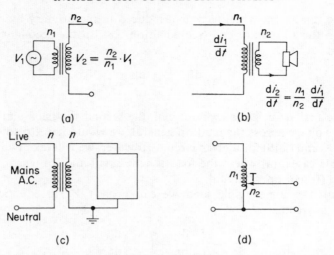

Figure 5.8. Types of transformer: (a) step–up voltage transformer, (b) step–up current transformer used with a loudspeaker, (c) mains isolating transformer and (d) autotransformer

circuit also acts as the secondary coil. In some autotransformers the tapping point T can be moved to any point along the coil, to give any desired turns ratio.

5.2 Inductive and Capacitive Circuits

We remarked earlier that a self–inductance in a circuit tended to oppose the change of current in the circuit (equation 5.11). For this reason, on applying a steady voltage across a circuit containing inductance and resistance in series the current takes a finite time to reach its steady value. On the other hand, if we apply a steady voltage to a circuit containing resistance and capacitance in series, it takes a finite time to charge the capacitor and current flows in the circuit until the voltage across the capacitor is equal to the applied voltage. In both cases the way in which the current approaches its final value (a maximum in the inductive circuit, zero in the capacitive one) is a similar function of time.

Consider first an inductive circuit (Figure 5.9a) in which a steady voltage V_0 is connected at a time $t = 0$. As in equation (5.11), the effective emf in the circuit is reduced by the back emf in the inductance, so we have

$$iR = V_0 - L\frac{di}{dt} \tag{5.22}$$

or

$$\frac{di}{dt} = \frac{R}{L}\left(\frac{V_0}{R} - i\right). \tag{5.23}$$

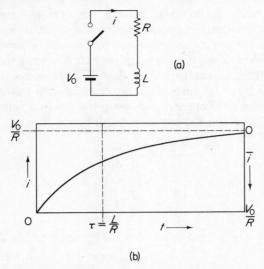

Figure 5.9.   The rise of current in an inductive circuit

In other words, the rate of increase of the current is proportional to the *difference* between its ultimate steady value, V_0/R (which it will attain when $di/dt = 0$), and the instantaneous value, i. When the difference is large (near $t = 0$) the current increases rapidly, but when the current nears its final value it approaches it more slowly (Figure 5.9b). Let the difference $(V_0/R - i) = \bar{i}$; an increase in i produces an equal decrease in $\bar{i}$, so that $di = -d\bar{i}$. We then have, from equation (5.23),

$$\frac{d\bar{i}}{dt} = -\frac{R}{L}\bar{i}. \tag{5.24}$$

The solution of the above differential equation is that

$$\bar{i} = \bar{i}_0\, e^{-(R/L)t}, \tag{5.25}$$

where $\bar{i}_0$ is the value of $\bar{i}$ at $t = 0$ (i.e. V_0/R). The difference $\bar{i}$ between the current and its final value thus decreases exponentially with time, until it becomes zero. In particular, we can see that after a time $t = L/R$, the difference becomes

$$\bar{i} = \bar{i}_0\, e^{-1} \simeq \bar{i}_0/3 \tag{5.26}$$

and has thus fallen to about a third of its initial value (Figure 5.9). The value $L/R = \tau$ is called the *time–constant* of the circuit as a whole, and is a measure of how long it takes the current to 'settle down'. (After three time–constants, the current is within e^{-3} or 5 per cent. of its final value.)

If, while a steady current was flowing, we were to remove the steady voltage from the inductive circuit (e.g. by short–circuiting the battery in Figure 5.9a), a current would continue to flow for a short time. This is because the work which is used to establish a current in an inductor is stored as the energy of the magnetic field produced by the current, and is released when the current and magnetic field are reduced. It is this energy which is available to prolong the current even when the external voltage is removed.

On removing voltage V_0, equation (5.22) becomes

$$iR = -L\frac{di}{dt}$$

or

$$\frac{di}{dt} = -\frac{R}{L}i. \tag{5.27}$$

The solution of this is the same as that of equation (5.24), that is

$$i = i_0\, e^{-(R/L)t}, \tag{5.28}$$

where now the circuit current itself decays exponentially from its initial value i_0 with a time constant $\tau = L/R$ (see Figure 5.10).

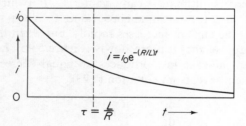

Figure 5.10.   Decay of current in an inductive circuit from which the applied emf has been removed

We can now treat the capacitive series circuit (Figure 5.11) in the same way as above. If we apply a steady voltage V_0 to a capacitance and resistance, the effective emf in the circuit is reduced by the voltage on the capacitor, so that

$$iR = V_0 - q/C, \tag{5.29}$$

where q is the charge on the capacitor and C its capacitance. Now $i = dq/dt$, the rate of flow of charge round the circuit, so that the equation becomes

$$\frac{dq}{dt} = \frac{1}{RC}(CV_0 - q), \tag{5.30}$$

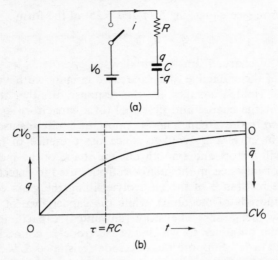

(a)

(b)

Figure 5.11. The rise of charge on a capacitor in series with a resistor

where CV_0 is the final charge on the capacitor. This is identical in form to equation (5.23), and the charge on the capacitor builds up in exactly the same way (Figure 5.11b) as the current in the inductance. Defining $\bar{q} = (CV_0 - q)$ we find

$$\bar{q} = CV_0 \, e^{-t/(RC)} \qquad (5.31)$$

and the time–constant of the capacitive circuit is $\tau = RC$. On discharging

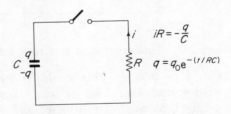

Figure 5.12. Discharging a capacitor through a resistance R

a capacitor through a resistance (Figure 5.12) there is no external voltage V_0 in the circuit, so that equation (5.29) becomes

$$iR = -q/C$$

or

$$\frac{dq}{dt} = -\left(\frac{1}{RC}\right)q, \qquad (5.32)$$

giving a solution (see equations 5.24 and 5.25) of the form

$$q = q_0 \ e^{-t/(RC)} \tag{5.33}$$

for the decay of charge from its initial value q_0.

The effect of the inductive and capacitive circuits we have discussed is to smooth out sudden changes, for the response of either the current (in an inductor) or the charge and potential (of a capacitor) lags behind any impulse applied to the circuit. This is important in circuits used to measure fast changes in current or voltage, since the presence of inductance or capacitance will distort and smooth out the shape of the measured signal waveforms. For instance, in the microelectrodes used to detect nerve action potentials, the resistance of the electrolyte filling the glass electrode may be tens or hundreds of megohms, while the capacitance of the electrode may be a few picofarads. The stray (unwanted) capacitance across the terminals of the amplifier to which the electrode is connected may also be of the same order. Suppose the electrode resistance R is 50 MΩ, while its capacitance is 6 pF, and the input capacitance of the amplifier is 4 pF (Figure 5.13). The time–constant, $\tau = RC$, of the detecting system is then

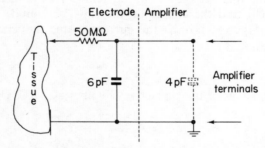

Figure 5.13.   Time–constant of a microelectrode and amplifier

$50 \times 10^6 \times 10 \times 10^{-12} = 0.50$ ms. This would be suitable for signals where the changes were spread over 5 ms or longer (i.e. $> 10 \tau$), but voltage 'spikes' lasting only 1 ms would be distorted and signals with a period of 0.05 ms or less would be smoothed out completely and remain undetected.

5.3 Alternating Current

The electromotive force produced by a dynamo or generator varies sinusoidally with time (equation 5.5) and the alternating current produced by the emf in a circuit also varies sinusoidally, although the peaks of current may not coincide in time with the peaks of voltage. We can represent the current i by an equation

$$i = i_0 \sin \omega t, \tag{5.34}$$

in which i_0 is the peak value of the current, t time and ω the *angular frequency* of the oscillations ($\omega = 2\pi f$, where f is the frequency in Hz). To express the fact that the alternating voltage V across a given circuit component may be out of step with the current, we write

$$V = V_0 \sin(\omega t + \phi), \tag{5.35}$$

in which ϕ, the *phase angle* added to ωt, shows how far (in angular terms) the graph of V is ahead of the graph of i. If ϕ is positive, the oscillations of V are ahead of those in i and the voltage is said to *lead* the current. If ϕ is negative, as in Figure 5.14, the voltage lags behind the current (in the figure,

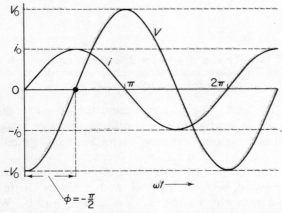

Figure 5.14. The phase angle between voltage and alternating current in a circuit component. $V = V_0 \sin(\omega t - \pi/2)$, $i = i_0 \sin \omega t$

by 90° or $\pi/2$ radians), so that peaks of V happen at a later time than peaks of i. Obviously when $\phi = \pm 2\pi n$ where n is a whole number, the phase shift of the oscillations with respect to each other is such that they again coincide, which is equivalent to $\phi = 0$.

5.3.1 *The Vector Diagram*

Instead of using equations like (5.35) to describe alternating quantities, we can also represent them graphically. A quantity $A = A_0 \sin \omega t$ can conventionally be represented by the height, or projection on a vertical axis, of a vector of length A_0 that rotates anticlockwise at a rate of ω radians per second, beginning from a position where it points horizontally to the right (Figure 5.15). The height of the tip of the vector is always given by $A_0 \sin \omega t$. Another quantity described by $B = B_0 \sin(\omega t + \phi)$ can similarly be represented by a vector of length B_0, rotating at the same rate (ω) as A_0 but always at an angle ϕ ahead of it. Provided all the quantities re–

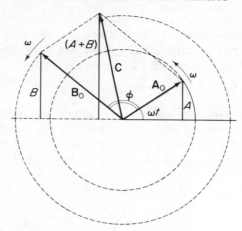

Figure 5.15. The use of a vector diagram to represent alternating quantities. The vector sum **C** of the vectors **A**$_0$ and **B**$_0$ has a vertical projection equal to $(A + B)$

presented have the same angular frequency, the corresponding vectors remain at fixed angles relative to one another, so we can imagine all the vectors being drawn on one common *vector diagram*, which rotates at ω rad s^{-1}.

It can also be shown that the sum of the oscillating quantities A and B is correctly represented, both in amplitude and phase, by the projection of a vector in the diagram which is the vector sum of **A**$_0$ and **B**$_0$. We can thus add and subtract alternating voltages, currents, etc., of different phases by very simple vector addition (i.e. drawing) instead of manipulating trigo–nometrical equations. Since phase difference must be measured *relative* to something, in A.C. theory the phases of quantities are expressed in relation to that of the circuit current. In the vector diagram, the vector **i**$_0$ is thus drawn horizontally ($\phi = 0$) and the phases of other quantities shown by the angles their vectors make with **i**$_0$.

5.3.2 *Average Values*

The mean value of a sinusoidally varying quantity is zero if the average is taken over a complete number of cycles, since the graph is symmetrical above and below the zero axis. A moving–coil meter (which cannot respond fast enough to follow oscillations at 50 Hz, and therefore reads an average current) would thus read zero if an alternating current were passed through it. Sometimes A.C. is measured by arranging that alternate half–cycles of current pass through the meter in the *same* direction, instead of opposite ones, so that the average current is not zero (Figure 5.16). The average current is the mean taken over half a period.

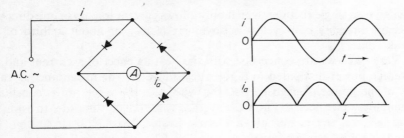

Figure 5.16. The use of rectifiers, which pass current in only one direction, to ensure that each half–cycle of current passes through a meter in the same direction

To calculate this mean, we note that the total charge passing through the meter in a half–cycle is the area under the graph of i against t (since $q = it$) for half a period, or for a graph of i versus ωt, $1/\omega$ times the corresponding area, which lies between $\omega t = 0$ and $\omega t = \pi$ (Figure 5.17). It can

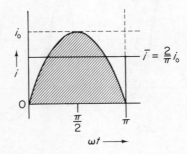

Figure 5.17. Average of the A.C. current over half a cycle. The shaded area is equal to $2 i_0$. The area in the rectangle of height $\bar{i}$ and length π has the same value

be shown that the area under half a cycle of a sine graph is twice the peak height, so in our case we have

$$\text{Total charge through meter} = \frac{1}{\omega}(2i_0),$$

$$\text{Time for half–cycle,} \qquad t = \pi/\omega,$$

$$\text{Average current } \bar{i} = \frac{\text{charge}}{\text{time}} = \frac{2}{\pi} i_0. \tag{5.36}$$

Thus the average current passed by the device shown in Figure 5.16 (a 'full–wave bridge') is about two–thirds of the peak current. The average current passed by a single rectifier, which only allows current to pass in one direction (i.e. on *alternate* half–cycles) is obviously $(1/\pi)i_0$, or about a third of the peak value.

Very often we are concerned with the heating effect of a current and the average power dissipated in a given resistance R. The instantaneous value W of the power is given by i^2R, but with A.C. the value of i is constantly changing, so we need to know the average value $(\overline{i^2})$ of i^2 in order to estimate, for instance, the power of an electric heater. Now $i^2 = i_0{}^2 \sin^2 \omega t$, but from trigonometry we have that $\sin^2 \theta = \frac{1}{2}(1 - \cos 2\theta)$, so

$$i^2 = i_0{}^2 \tfrac{1}{2}(1 - \cos 2\omega t). \tag{5.37}$$

A graph of i^2 is shown in Figure 5.18, and from this or the above equation it can be seen that the effect of squaring the alternating quantity is to produce a value which still varies sinusoidally, but at *twice* the frequency (i.e. we

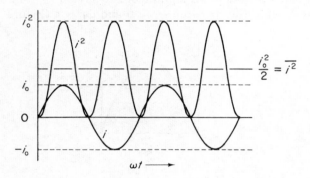

Figure 5.18.   A graph of i^2 versus ωt to show (a) frequency doubling and (b) the value of $\overline{i^2}$

obtain a *second harmonic* of the original frequency). The average value of i^2 is the level about which the i^2 graph is symmetrical (Figure 5.18), so that

$$\overline{i^2} = i_0{}^2/2, \tag{5.38}$$

a result which also follows if we average both sides of equation (5.37) over one or more cycles. The same treatment would, of course, give $\overline{V^2} = V_0{}^2/2$, and so on for other alternating quantities; the 'mean–square' value is half the square of the peak value.

If we take the square root of the mean–square value of, for example, current, we obtain an effective value of current which is called the root–mean–square value (RMS). Thus,

$$i_{RMS} = \sqrt{\overline{i^2}} = \sqrt{i_0{}^2/2} = i_0/\sqrt{2}. \tag{5.39}$$

The average power $\overline{W}$ developed in a resistance is given by

$$\overline{W} = \overline{i^2}\, R,$$

$$\therefore \quad \overline{W} = i_{RMS}{}^2\, R, \tag{5.40}$$

whereas under D.C. conditions we simply have that $W = i^2 R$. The value of i_{RMS} is therefore the equivalent D.C. value of the alternating current, as far as power is concerned; we can use i_{RMS} in the D.C. power equation and obtain the correct value for the power in the A.C. situation. Similarly, $V_{RMS} = V_0/\sqrt{2}$ is the equivalent D.C. voltage, for

$$\overline{W} = V_{RMS}{}^2/R. \tag{5.41}$$

Practically, alternating currents and voltages are usually specified by giving their RMS values, *not* the peak ones. For instance, a mains voltage of 250 V A.C. implies that $V_0/\sqrt{2} = 250$, $V_0 \simeq 350$ V. This is a point to bear in mind when insulating mains–operated devices in which, at a given instant, one part may be at $+V_0$ and another at $-V_0$, a potential difference of 700 V in this case.

5.3.3 *Measurement of A.C.*

We have already seen that ordinary moving–coil meters can be adapted to register alternating current (Figure 5.16) by the use of rectifying devices (Chapter 6) which pass current in one direction only. Naturally the meter scale must be specially calibrated, since the pointer movement is proportional to mean current ($\dfrac{2}{\pi} i_0$ or $\dfrac{1}{\pi} i_0$) whereas we wish the scale to read in RMS values ($i_0/\sqrt{2}$). The finite resistance of the rectifiers has also to be taken into account.

Other meters exist that give a deflection that depends on i^2, which is always positive, so they can be used equally well with A.C. or D.C. For instance, in the *thermocouple ammeter* the current flowing in a wire raises its temperature by an amount which depends on the Joule–heating effect of the current. The hot wire has a thermocouple junction attached to the mid–point, and the thermocouple current (which depends on the wire temperature) is registered by an ordinary D.C. meter.

In the *dynamometer* magnetic forces are used. Current i_1 flows in a solenoid coil and produces a magnetic field. In the field a smaller coil is suspended, carrying a current i_2. As in the moving–coil meter (Chapter 4), the field exerts a couple on the smaller coil, which is registered by a pointer. In this case the field is proportional to i_1 and the couple $G \propto i_1 i_2$. If the same current i is arranged to flow in both coils, $G \propto i^2$ and the system can be used as an A.C. meter. A more interesting use of the device is to arrange that the current i_1 in the 'field' coil is proportional to the current I flowing through a load, while i_2 is proportional to the voltage V across the load

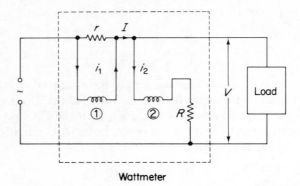

Wattmeter

Figure 5.19.   Use of the dynamometer as a wattmeter to measure power. Coil 2 is subject to a torque in the field produced by coil 1, which is stationary. R is a high resistance

(Figure 5.19). Then $G \propto i_1 i_2 \propto IV$ and we have a *wattmeter*, since IV is the power developed in the load, in watts. The wattmeter works for either A.C. or D.C. circuits.

Other instruments for making A.C. measurements are the *'valve' volt–meter* (which now uses transistors) and the *cathode–ray oscilloscope*. The latter is described in Chapter 6.

5.3.4   A.C. in Reactive Components

If we apply an alternating voltage to a capacitor (Figure 5.20), positive charge will first build up on one plate of the capacitor and then flow round

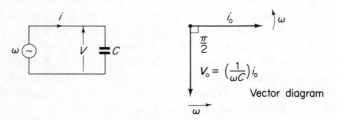

Figure 5.20.   The capacitor in an A.C. circuit

the circuit to build up on the opposite plate—an alternating current will flow in the circuit. If V is the applied voltage, the charge on the capacitor is $Q = CV$ while the current flowing to the capacitor is given by $i = \mathrm{d}Q/\mathrm{d}t = C\,\mathrm{d}V/\mathrm{d}t$. Suppose that $V = V_0 \sin \omega t$, then

$$\frac{\mathrm{d}V}{\mathrm{d}t} = \omega V_0 \cos \omega t,$$

$$\therefore \quad i = C\frac{dV}{dt} = [\omega C \; V_0] \sin(\omega t + \frac{\pi}{2}), \tag{5.42}$$

since $\cos \alpha = \sin(\alpha + 90°)$. We see that the current oscillations in the circuit are ahead of the voltage oscillations in phase by $\pi/2$ or $90°$. Since we take current as the standard, it is more conventional to say that the phase of the voltage *lags* $\pi/2$ behind the current, as shown in Figure 5.14. In equation (5.42) the quantity in square brackets is the peak current, i_0, so that

$$i_0 = \omega C \; V_0$$

or

$$\left(\frac{1}{\omega C}\right)i_0 = V_0. \tag{5.43}$$

Comparing the above equation with $RI = V$ for the current in a resistor, it is evident that $1/(\omega C)$ is an analogous quantity to R. This quantity is called the *reactance*, X, of the capacitor and represents the opposition of the component to the passage of current. Since $X = 1/(\omega C)$, the reactance falls with increasing frequency, so that a capacitance passes high–frequency currents easily, whereas a direct current ($\omega = 0$) meets an infinite reactance and is unable to pass.

In the vector representation of voltage and current (Figure 5.20), the vector i_0 is drawn horizontal and the vector $\mathbf{V}_0$ is drawn downwards, since V in a capacitor is *behind* i by $90°$ and the diagram is imagined to rotate anticlockwise.

In a similar way we can investigate the current in an inductor (Figure 5.21). The instantaneous current i in the inductance L is given by $V = L \, di/dt$

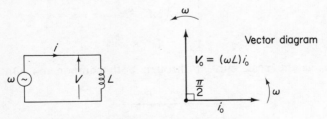

Figure 5.21.   The inductor in an A.C. circuit

(assuming the resistance of the real inductor coil to be negligible). If $i = i_0 \sin \omega t$ we have

$$\frac{di}{dt} = \omega i_0 \cos \omega t,$$

$$\therefore \ V = L\frac{di}{dt} = [\omega L \, i_0] \sin(\omega t + \frac{\pi}{2}), \tag{5.44}$$

and here the voltage is *ahead* of the current by a phase angle $\pi/2$ radians. (In other words, in equation 5.35, $\phi = +\pi/2$.) Since the peak voltage $V_0 = \omega L \, i_0$, the reactance of the inductor is given by

$$X = \frac{V_0}{i_0} = \omega L \tag{5.45}$$

and, as we might expect from the nature of inductance, the reactance *increases* for currents of higher frequency, which vary more rapidly. An inductor will thus pass direct current easily, but oppose the passage of high–frequency A.C.

Because the voltage leads the current, the voltage vector (Figure 5.21) is drawn upwards, $\pi/2$ in front of the current vector.

If we have an A.C. series circuit containing, for example, a resistor and capacitor (Figure 5.22), we can use the vector diagram to add the alternating

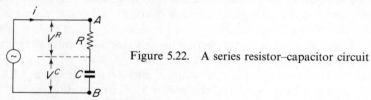

Figure 5.22.   A series resistor–capacitor circuit

voltages that appear across each component, and find the total voltage between points A and B. If V_0^C is the peak voltage across C and V_0^R that across R, we have

$$V_0^C = \left(\frac{1}{\omega C}\right) i_0$$

and

$$V_0^R = R \, i_0,$$

where i_0 is the peak current through both components. On the vector

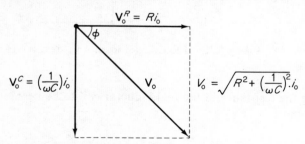

Figure 5.23.   The vector diagram for the circuit of Figure 5.22

diagram (Figure 5.23) $\mathbf{V}_0^R$ is drawn horizontally (since, in a resistor, voltage is just proportional to current and in step with it) while $\mathbf{V}_0^C$ is drawn $90°$ behind. Combining the two vectors, the resultant (vector sum) is represented by the vector $\mathbf{V}_0$, which represents the sum of the alternating voltages across C and R (i.e. the voltage between A and B). From the right–angled triangle of Figure 5.23 we have

$$V_0{}^2 = (i_0\,R)^2 + \left(i_0\,\frac{1}{\omega C}\right)^2, \tag{5.46}$$

$$V_0{}^2 = i_0{}^2\left[R^2 + \left(\frac{1}{\omega C}\right)^2\right], \tag{5.47}$$

or if

$$V_0 = i_0\,Z, \qquad Z = \sqrt{R^2 + \left(\frac{1}{\omega C}\right)^2}. \tag{5.48}$$

Here Z, the ratio between the current and the total voltage applied across AB, is called the *impedance* of the circuit between A and B and includes the effect of both the resistance (R) and the reactance $1/(\omega C)$ in the circuit. More generally, if we have a total resistance R in series with a total reactance X, the impedance of the combined components is given by

$$Z = \sqrt{R^2 + X^2}. \tag{5.49}$$

If, to Figure 5.22, we add an inductance in series with the capacitance, it can be shown that the total reactance of the circuit is the *difference* of the inductive and capacitive contributions, so that

$$X = \left(\omega L - \frac{1}{\omega C}\right). \tag{5.50}$$

Such a circuit then exhibits *resonance*, for at one particular value of ω, X^2 drops to zero and the impedance of the whole circuit (equation 5.49) falls to the minimum value of $Z = R$. At the resonant frequency at which this occurs, current in the circuit rises to a maximum.

5.3.5 *The Transformer and Impedance Matching*

We have already seen (Section 5.1.3) how a transformer produces an emf in the secondary coil when the current in the primary circuit varies. If the primary current is A.C., the secondary emf and current are also alternating. Depending on the circuit it is in, the transformer acts either as a current or voltage transformer (see equations 5.20 and 5.21). Let us take, for example, the current transformer as shown in Figure 5.24a. The average power in the secondary load resistance R_s is $i^2\,R_s$, and, if we assume that power dissipation in the transformer itself is negligible, the primary power input

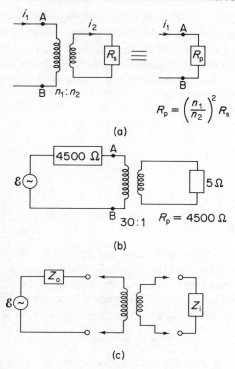

Figure 5.24. The apparent resistance presented by a transformer.
(a) Theory. (b) Use to obtain maximum power transfer to a load.
(c) Impedance matching in general

must be the same (since energy is conserved). If R_p is the apparent resistance between the points AB, we therefore have

$$\overline{i_1^2}\, R_p = \overline{i_2^2}\, R_s, \tag{5.51}$$

but

$$i_2 = \left(\frac{n_1}{n_2}\right) i_1 \quad \text{if } R_s \to 0,$$

$$\therefore \quad R_p = \left(\frac{n_1}{n_2}\right)^2 R_s. \tag{5.52}$$

The effect of the transformer is to present, across AB, an effective load which is the secondary load multiplied by the square of the turns ratio—the secondary load is said to be 'reflected' into the primary circuit. (If we took a voltage transformer and computed the power as V^2/R, we would obtain the same effect.) As we saw in Chapter 4, maximum power is trans–ferred to a load if it is equal or 'matched' to the internal resistance of a generator. A load can therefore be matched to the internal resistance of an A.C. generator by interposing a transformer of appropriate turns ratio between the two (e.g. as in Figure 5.24b).

It can be shown that a transformer 'reflects' the reactance of a secondary circuit into the primary, as well as the resistance. More generally, therefore, the transformer can be used to match the impedance of two parts of a circuit that are to be connected. The part of the circuit that provides the emf is said to have an 'output impedance', Z_o, analogous to the internal resistance of a generator, while the other part of the circuit has an 'input impedance', Z_i, analogous to a load (Figure 5.24c). Optimum power transfer is obtained if Z_i is matched, directly or indirectly, to Z_o.

5.4 Electromagnetic Radiation

We have seen in Chapter 4 that a steady current (charge moving with constant velocity) produces a steady magnetic field. Suppose, however, we have the arrangement of Figure 5.25 in which charges are alternately ac–

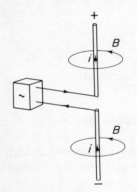

Figure 5.25.   A dipole rod aerial which produces electromagnetic radiation

celerated from the end of one rod to the end of another. (The two rods are equivalent to a dipole, with opposite charges at each end which continually reverse in sign; the device is called an oscillating dipole.) The accelerating charges will produce a changing magnetic field around the rod. At the beginning of this chapter, however, we mentioned that a changing magnetic field produces an electric field, able to induce currents in conductors.

The oscillating current going up and down the rods thus produces an oscillatory magnetic field around the dipole, which in turn gives rise to an oscillating electric field. If this were all that happened, and the electric field had no further effect, it can be shown that no energy would be lost from the dipole (apart, of course, from Joule heating of the rods). However, Maxwell, who was investigating the relations between electric and magnetic fields, showed theoretically that a changing electric field (even in space, where it gives rise to no conduction current) has the *effect* of a current, in that it produces a magnetic field. (The effective current is called a displacement current, since it exists when the displacement, D ($= \varepsilon E$), is changing with time.) In other words, an oscillating electric field produces an oscillating magnetic field.

Thus, around our dipole, the alternating electric field produces an alternating magnetic field, which produces an alternating electric field, which produces ..., etc. The association of the two fields in this way is the necessary condition, as Maxwell showed, that *waves* of varying electric and magnetic fields should be propagated through space. The oscillating dipole does, in fact, radiate energy out into the space around it, in the form of *electromagnetic waves*. At a point in space which receives the radiation there is both an electric field and a magnetic field directed at right angles to one another. Both fields are at right angles to the direction of travel of the radiation, as shown in Figure 5.26, and for this reason the radiation is said

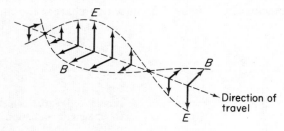

Figure 5.26. The transverse nature of electromagnetic radiation

to be *transverse*. In the case of the oscillating dipole the radiated electric field vector always lies in the same plane as the dipole axis, and the radiation is said to be *plane–polarized*.

Purely from a consideration of the electromagnetic equations, Maxwell was able to predict that the velocity of the radiation in a vacuum should be given by

$$c = \frac{1}{\sqrt{\varepsilon_0 \, \mu_0}}. \tag{5.53}$$

Here ε_0 and μ_0, the permittivity and permeability of space (Chapter 4), are measurable values and give a value for c equal to the velocity of light, $3{\cdot}00 \times 10^8$ m s^{-1}, in a vacuum. It is now an accepted fact that radio waves, light, X–rays and γ–rays are all electromagnetic radiation and differ only in their frequency and wavelength, though all have the common velocity c in free space.

Whenever an electron is accelerated (or decelerated, which is acceleration in the reverse direction to the velocity), electromagnetic radiation is produced.

Apart from the radio dipole aerial which we considered (Figure 5.25), the same phenomenon occurs, for instance, when the oscillating charges in a vibrating molecule radiate infrared light. Another instance is in the production of X–rays. When an electron travelling at high velocity collides with matter and is violently decelerated, radiation is produced of correspondingly high energy. In an X–ray generator part of the output of the machine arises in this way, but even in a colour television set the electrons hitting the face of the tube have enough energy to produce 'soft' X–rays.

Further Reading

Duffin, W.J. (1965). *Electricity and Magnetism*. McGraw–Hill, New York.
Kip, A.F. (1962). *Fundamentals of Electricity and Magnetism*. McGraw–Hill, New York.
Richards, J.A., F.W. Sears, M.R. Wehr and M.W. Zemansky (1960). *Modern University Physics*, Part 2. Addison–Wesley, Reading, Mass.

PROBLEMS

5.1 A coil of 1000 turns and cross–sectional area 2 cm^2 lies in a uniform magnetic field so that the plane of its cross–section is normal to the field lines (Figure 5.2). If the induction in the field is 0·5 T, calculate (a) the total flux through the coil and (b) the maximum emf induced in it when it is rotating at 3000 revolutions per minute.

5.2 Calculate the back emf developed in an inductance of 0·3 H when the current through it changes at 1 A per millisecond.

5.3 A signal amplifier has an input impedance equivalent to a capacitance of 0·0001 μF and a resistance of 2·5 MΩ connected in parallel between the input terminals. What is the time–constant of the system? How would the amplifier respond to sinusoidally varying signals of frequency (a) 200 Hz, (b) 4 kHz, (c) 0·1 MHz?

5.4 The measured RMS current through a heating coil is 2·0 A while the peak voltage across the coil is found to be 354 V by displaying the waveform on a cathode–ray oscilloscope. At what power does the heater operate?

5.5 A circuit comprises a capacitance, an inductance of 1 H and a resistance of 50 Ω connected in series across an A.C. supply of frequency 1000 Hz. Calculate (a) the reactance of the inductor, (b) the value of capacitance necessary to make the circuit become resonant at the supply frequency and (c) the RMS current in the circuit at resonance if the supply voltage is 5 V$_{RMS}$.

5.6 In the circuit of problem 5 above, using the value of capacitance needed to give resonance at 1000 Hz, find the reactances of the inductor and capacitor if the supply frequency were reduced to 800 Hz. What, then, would be the total impedance of the circuit?

5.7 Find the turns ratio (primary : secondary) of a transformer that will match a loudspeaker of nominal impedance 15 Ω and a thermionic–valve amplifier with an output impedance of 24 kΩ.

CHAPTER 6

Electronics

Electronics can be loosely defined as a technology based on the control of electron flow by vacuum or solid–state devices. Except in special cases the purpose of this control is to obtain, convey or display information, and it is for this reason that electronics has become such an important tool in medicine and biology.

The early development of electronics was based on the use of the vacuum tube or 'radio valve' in which the source of electrons (the cathode) is a heated conductor, from which the electrons escape by means of their thermal energies. In such thermionic valves the flow of the electrons through the vacuum towards a positive electrode (the anode) is controlled by the presence of electric fields due to other electrodes ('grids') interposed between the cathode and anode. This important principle, that the amount of current flowing between anode and cathode can be controlled by quite small potentials applied to the grids, was applied with great ingenuity in the development of radio transmitters and receivers, amplifiers, the electron microscope (Chapter 11), the cathode–ray tube as used in television, and all the other electronic devices with which we are now familiar.

In recent years, however, other methods have been found for manipulating the flow of electrons, which depend on the use of semiconducting materials, for instance in the transistor. In such a solid–state device electrons are, of course, already present, and the ways in which they may be induced to flow through the solid constitute methods of controlling a current in the device. Solid–state devices have, in fact, proved so successful that they have made the thermionic tube almost obsolete except for special applications in which it excels.

Table 6.1 compares the features of the two systems, which may in some cases be either advantages or disadvantages, depending on the application in view. For instance, the temperature–dependent behaviour of semiconductors is generally a disadvantage, but if one wishes to measure temperature it becomes an advantage and the semiconducting thermistor (Chapter 2) has wide uses as a small temperature sensor. The small size, reliability and low power requirements of transistors make them ideal for

TABLE 6.1

A comparison of the features of vacuum–tube and solid–state devices in electronics.
Rough orders of magnitude are given in parenthesis

Vacuum	Solid–state
Medium size (cm)	Small size (mm)
Limited life inherent in the thermionic system (1000 hr)	Reliable in proper use, with effectively infinite life
Electrically rugged	Easily destroyed by improper electrical conditions
Needs, and can handle, high voltages (> 100 V)	A low–voltage device (10 V)
High input and output impedance	Low impedance system (except, for example, field–effect transistors)
High power handling capacity	Medium power capacity (W)
Controls small currents (mA)	Can control large currents (A)
Behaviour insensitive to ambient temperature	Behaviour depends markedly on temperature

many biological applications. On the other hand, when biological potentials are to be detected (e.g. in neurophysiology, electroencephalography, etc.) a high input impedance is necessary for the detecting system and vacuum tubes of special design are generally used. Even here, field–effect transistors (FETs) are available which challenge the performance of the vacuum tube.

In this chapter we shall deal only briefly with the thermionic valve, and then describe other vacuum devices, such as the cathode–ray oscilloscope and photomultiplier, which are in common use. As befits their importance, solid–state devices will receive greater attention.

6.1 Thermionic Vacuum Tubes

The electrons responsible for conduction in a metal, although free to move within the conductor, cannot escape from the surface unless they acquire a certain minimum energy, called the *work function* of the material. On heating the conductor, either directly (by passing a current through it) or indirectly (with an electric heater), the electrons are given the necessary thermal energy to escape and *thermionic emission* occurs. The cathode of a vacuum tube consists of a directly heated tungsten filament, or of an indirectly heated electrode coated with alkaline–earth oxides. The latter has the advantage of a lower work function and consequently a lower working temperature. It can also be electrically insulated from the heating current, which then does not interfere with the rest of the circuit.

The simplest vacuum tube is a two–electrode system or *diode* (Figure 6.1), in which the positive electrode (anode) is a cylinder surrounding the cathode. If the anode is made slightly positive with respect to the hot cathode, elec–

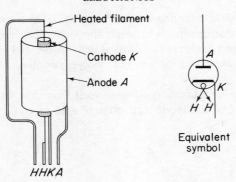

Figure 6.1.   An indirectly–heated diode valve (the glass envelope is not shown)

trons are attracted to the anode and a current flows through the valve. (The conventional current is from anode to cathode, since electrons carry negative charge.) Unless electrons can be pulled away from the cathode as fast as they are emitted, a negative cloud of electrons (a 'space charge') builds up in front of it, which tends to prevent the current flow. In this situation the current that does flow increases with the anode potential. However, above a certain anode potential the current 'saturates' and reaches its maximum value, which is dictated by the temperature of the cathode (Figure 6.2).

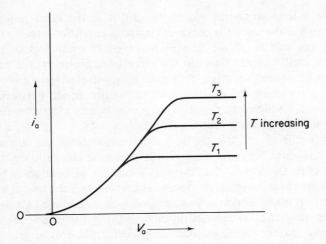

Figure 6.2.   The current (i_a) between anode and cathode in a diode valve, as a function of anode–cathode potential difference (V_a) and cathode temperature (T)

If the anode is made more *negative* than the cathode, it repels the thermionic electrons and no current flows through the valve. The diode therefore has important rectifying properties, since it will allow only one direction of current flow. The use of rectifiers in D.C. power supplies is treated in Section 6.5.1.

The *triode valve* (Figure 6.3) is developed from the diode by having a loosely–wound spiral electrode, the 'control grid', around the cathode and

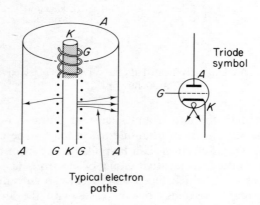

Figure 6.3.   Detail of the triode valve construction, showing the control grid

nearer to it than to the anode. If the grid is at the same potential as the cathode, while the anode is positive, electrons can flow between the loops of the grid and still reach the anode. (Some electrons in fact strike the grid, so that a small current flows in the circuit connected to the grid.) As the grid is made more negative, it begins to repel electrons and restrict the numbers that can get from the cathode to the anode between its loops. If it is made sufficiently negative, it prevents any electrons reaching the anode and the anode current is completely cut off.

The potential V_g of the control grid thus controls the current through the valve (Figure 6.4), and since the grid is near the cathode it has much more effect on the current than the anode does. Small changes in the negative grid voltage cause changes in the anode current, which it would take much larger variations in anode voltage to produce. The valve can thus act as a voltage amplifying system. Additionally, the grid, when negative, takes hardly any current itself, so that the power input to the grid is minimal while the power which it controls through the valve (anode voltage × anode current) may be of the order of several watts. The triode is thus capable of power amplification.

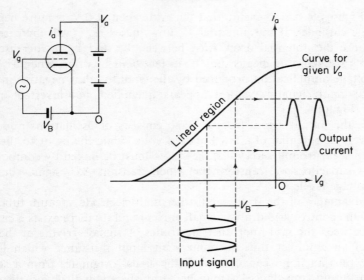

Figure 6.4.   Variation of anode current (i_a) with grid voltage (V_g) in
a triode circuit

In an amplifier, several valves are used in successive stages, so that the output signal from one is applied as the input of the next. The amplitude of the signal as it passes through the system is thus multiplied successively by the amplifying factor or 'stage gain' associated with each stage. An example of a voltage–amplifying stage is shown in Figure 6.5. Here a D.C.

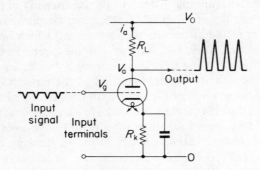

Figure 6.5.   Voltage amplification with a triode valve

potential V_0 of a few hundred volts is connected to the anode through a 'load resistor' R_L. The anode current i_a passes through R_L and the valve and through a 'bias resistor' R_k which has the same purpose as the battery

V_B in Figure 6.4 (i.e. to ensure that the grid, when at 0 V, is more negative than the cathode). If the potential V_g now increases, i_a will increase and thus cause the potential drop, $i_a R_L$, between V_0 and V_a to increase too. Since V_0 is fixed, this means that V_a in fact *falls* in value, when V_g rises. The voltage amplification produced by the circuit is thus negative in sign, and the input signal waveform appears magnified, but inverted, at the output (Figure 6.5).

Since the output signal from a stage consists of oscillations about an average D.C. potential of a few hundred volts, while the input to the next stage must be around zero volts, the stages must be linked by components such as capacitors or transformers, which transmit A.C. signals but not D.C. voltage levels.

Disadvantages of the triode and other multielectrode vacuum tubes are that (a) the control grid draws a small current and (b) there exists a capacitance between the grid and other electrodes. A signal arriving at the first valve of an amplifier thus encounters an input resistance which is not infinite, plus an input capacitance. If the signal originates from a source of high output impedance, it may be seriously attenuated and distorted (see Chapter 5). For this application, special valves are designed (electrometer triodes) which have input resistances as high as 10^{13} Ω and interelectrode capacitances of only fractions of a pF.

6.2 The Cathode-ray Oscilloscope

The cathode–ray oscilloscope (C.R.O.) and the television receiver are basically the same instrument. In the television tube the beam of electrons ('cathode rays') which strikes the phosphorescent screen is continually swept across it in a regular scanning system, while the intensity of the beam varies to produce the picture. In the C.R.O. the beam is stationary until signals are applied (from inside or outside the instrument) which have the effect of deflecting the beam, while the intensity of the beam remains constant. The beam can thus be made to draw a uniformly bright line on the screen, which remains visible until the light emitted by the phosphorescent coating dies away. If the beam retraces the same path continually at high frequency, the path remains permanently visible and appears stationary.

A schematic diagram of the tube of a C.R.O. is shown in Figure 6.6. Electrons from a heated cathode K are attracted through a small hole in the end of a cylinder G towards an anode A_1 which may be several kV more positive than K. G is negative with respect to K and tends to prevent electrons passing through the hole—variations in the potential of G therefore control the intensity of the electron beam. After passing through the aperture of A_1 the electron beam is focused by passing through the electric field created by A_1, A_3 and the cylinder A_2, which together act as an electrostatic lens and ensure that the electrons strike the screen S within a small

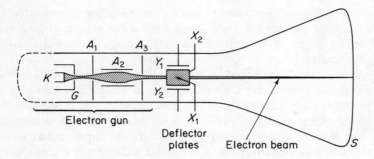

Figure 6.6.   Schematic diagram of the cathode–ray oscilloscope. (K, G and A_2 are cylinders; A_1 and A_3 are discs; plate X_2 is behind X_1.)

spot. (The complete electrode system from K to A_3 is called the *electron gun.*)

To move the focused spot across the screen, electric fields are used to deflect the beam issuing from the electron gun. For example, a potential difference between the plates Y_1 and Y_2 creates a vertical electric field which deflects the beam and the spot in the vertical (Y) plane. Similarly, a voltage signal applied across the X plates will move the spot horizontally, in the 'X'–direction. The X and Y plates are 'driven' by amplifiers within the instrument, so that quite small input signals (of the order of 1 mV or less, depending on the amplifier) can produce reasonable deflections on the screen.

Generally, we wish to display on the screen a 'graph' of a given input signal versus time. To do this, the signal is applied to the Y amplifier while the X amplifier is connected, within the oscilloscope, to a 'time–base circuit' which generates a saw–tooth voltage waveform (Figure 6.7). The rising

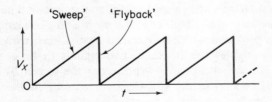

Figure 6.7.   The saw–tooth voltage waveform applied to the X plates of a C.R.O.

voltage of the saw–tooth (the 'sweep' voltage) moves the spot across the screen in the X–direction at a constant speed, so that the X–coordinate of the spot is proportional to time. For example, 1 cm may correspond to 5 s or, at the other extreme, to 5 μs. At the same time, the Y–coordinate depends on the input signal and the spot thus traces out the required graph. When the spot reaches the right–hand side of the screen, it then quickly flies back to its initial X–position (due to the sharp drop of the saw–tooth voltage).

When the time–base circuit is 'free–running', the sweep and flyback of the spot follow each other continuously as in Figure 6.7. However, if we wish to display a repetitive signal (such as a sine wave), it is convenient to prevent the sweep beginning until the input signal reaches a certain voltage level. The time–base is then *synchronized* with the input signal, as shown in Figure 6.8, and the same part of the signal waveform is traced out by the

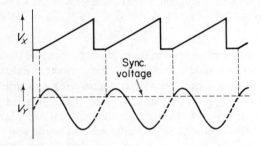

Figure 6.8. Synchronization of the time–base waveform with the signal applied to the Y plates of a C.R.O. Only the solid parts of the sine curve are displayed on the screen

spot on each sweep, so that we obtain an apparently stationary pattern on the screen. Another way of using the time–base is to prevent the sweep occurring until it is *'triggered'* by some externally applied signal such as a voltage pulse. For instance, in displaying the effects of electrically stimulating a nerve, we could trigger the sweep to begin at the precise time of stimulation.

Although we have described a C.R.O. with a single beam, very often an oscilloscope is provided with two electron beams, each with its own pair of Y plates. This enables the traces of two signals to be compared simul–taneously, as the two spots are swept across the screen by the time–base. Another useful version of the C.R.O. is the *storage oscilloscope*, in which a trace made on the screen can be retained if necessary for several hours, for measurement and photography, and can then be erased at the touch of a button. For comparison, conventional cathode–ray screens which are coated with even the 'slowest' phosphors will only display a single trace for a few seconds.

6.3 The Photomultiplier

The photomultiplier is basically a vacuum tube in which the source of electrons is photoelectric instead of thermionic; it is used as a very sensitive light detector. The light falls on a *photocathode K* (Figure 6.9), which is coated with a material whose work–function is small enough to allow electrons to be released by the light energy. (See Chapter 7 for more details

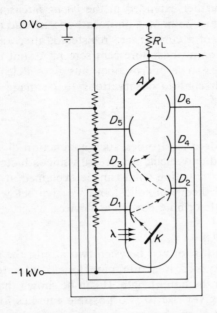

Figure 6.9. A diagram of one type of photomultiplier with its asso–ciated chain of resistors. λ denotes incident light

of this photoelectric effect.) In a simple photocell the small electron current is collected directly by a single anode, but in the photomultiplier the electrons from K are attracted to D_1, the first of a number of electrodes called *dynodes* whose potentials increase in sequence from a few hundred volts up to a few thousand.

An electron that has accelerated towards D_1 strikes it with enough kinetic energy to release two or more 'secondary' electrons from the dynode surface. These secondary electrons are in turn attracted to the dynode D_2, which is at a higher potential than D_1, and there give rise to further secondary emission. The process is repeated from one dynode to the next, so that a single electron from the photocathode may (depending on operating conditions) cause the emission from the final dynode of as many as 10^6 secondary electrons, which are collected by the anode A. Compared with the simple photocell, the photomultiplier may thus have a current amplification factor of about a million. The anode current, flowing through a fairly high resistance, produces a voltage across the resistor, R_L, which can be further amplified if necessary.

The photomultiplier is as good as the eye in detecting very weak light signals, and has a wavelength sensitivity (see Chapter 7) which covers ultraviolet, visible and near infrared light. The principle of photoelectric detection has been further extended in the *image intensifier*, in which the electrons released from a point on a photocathode, instead of being collected together in the form of a current, are registered, after amplification, at a corresponding point on a phosphorescent screen. A faint image formed on the cathode surface (e.g. by a microscope) thus gives a clearly visible image on the screen, which enables observations to be made in adverse light conditions.

6.4 Solid-state Devices

The devices of modern electronics, such as junction diodes, thermistors, transistors, photoconductive cells, etc., are all semiconductors. The electrical behaviour of a semiconducting solid can be explained at the atomic level by the electron–band theory of solids, and we give below a brief account of the theory before describing particular devices.

6.4.1 *Electrons in Solids*

When a stable bond is formed between two atoms, the potential energy of the molecule so formed is a minimum for an interatomic separation which corresponds to the bond length. It can be shown that such an interaction of two atoms gives rise to two possible energies for an electron of the molecule for each allowed energy level in the single atom. The atomic energy level is said to exhibit twofold degeneracy (Figure 6.10), as the graph

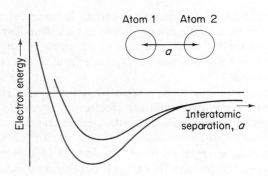

Figure 6.10. Twofold degeneracy in the energy level associated with an atomic electron. The two levels arise when two atoms interact to form a molecule

of electron energy against interatomic distance splits into two curves as the atoms approach each other. In a similar way, the interaction of N atoms (e.g. in a crystal) leads to N–fold degeneracy of the energy levels, so that in a solid containing a large number of atoms there will be very many energy levels. However, the energy difference between the lowest and highest energy level in a set derived from a single atomic level is little different whether two or a million atoms are concerned. Thus, the more energy levels there are in the set, the more closely spaced in energy they become.

Figure 6.11a gives an impression of the multifold degeneracy that would arise in the formation of a solid from individual atoms. If the equilibrium

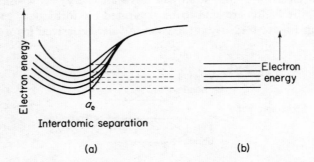

(a)   (b)

Figure 6.11.   Illustrating the formation of an energy band (b) from the energy levels which arise when many atoms or molecules combine to form a solid. a_e represents the equilibrium separation of neighbouring atoms or molecules. In (b) only the separation of the lines has any significance

interatomic spacing in the solid is known, an *energy band* can be drawn for the material (Figure 6.11b). Each level in the band represents a possible energy for an electron in the solid. The total 'width' of a band can be of the order of 500×10^{-21} J. (For comparison the thermal energy of atoms at room temperature is c. 4×10^{-21} J.) At the same time, a typical solid might contain 10^{23} atoms and give 10^{23} levels within the band, separated therefore by minute increments in energy of the order of 10^{-41} J.

The significance of this band structure lies not in the relative energies of the electrons within a band but in the fact that a particular band can hold only a limited number of electrons. This is because the Pauli exclusion principle allows only two electrons to occupy a given level within the band. The differences between conductors and insulators can be explained by the ways in which the bands are filled with electrons.

Consider a solid containing N atoms, so that an energy band has N levels. Since each level can hold two electrons, the whole band can hold $2N$ electrons. Now the energy bands, and the electrons which are available to fill them, derive from energy levels in the atom. Therefore, if an energy level of a single atom is normally full (containing two electrons) that level will provide $2N$ electrons in the solid, which just fill the corresponding energy band. The lower energy levels in atoms are normally fully occupied, and give rise to narrow bands in the solid which, for the above reason, are usually completely full. Atomic energy levels that contain only one electron give rise to half–filled energy bands. In many cases, bands derived from adjacent energy levels of the atom actually overlap in energy in the solid, to form a single, wider, band which may be partially occupied to an extent which differs from a half.

Since no electron in the solid can have an energy which lies between two bands, the bands are said to be separated by 'forbidden' energy gaps, as shown in Figure 6.12. Here the *valence band* is derived from the higher

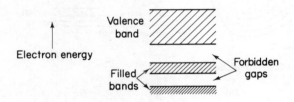

Figure 6.12.   Electron–band structure of a solid

energy levels of the atom and may or may not be completely full. If it is full the material is an *insulator* or *semiconductor*, while if it is only partially occupied the material is a *conductor*. For an electron to be able to move freely through the solid (the process of conduction), it must be raised to an energy level above the highest one that is occupied in the minimum–energy or 'ground' state of the solid. Electrons that have been raised or 'excited' to such a higher level are said to occupy the *conduction* energy band and will move through a solid under the influence of an electric field, thus constituting a current.

In a conductor (Figure 6.13a), where the valence band is only partially full in the ground state, the separation of levels within the band is so minute that the electrons can very easily move to higher energy levels within the same band, and the conduction band and valence band thus coincide. However, when the valence band is completely full, as in a semiconductor or insulator, the next highest energy level is in another, higher, energy band so that the conduction band is separated from the valence band by an energy gap (Figures 6.13b and c). In an insulator the gap is so large that few electrons acquire enough thermal energy to reach the conduction band and the conductivity of the solid is very small. In a semiconductor the gap

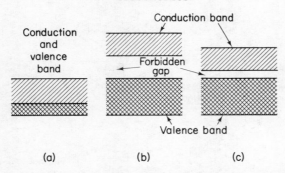

Figure 6.13. Occupation of electron energy bands in (a) a conductor, (b) an insulator and (c) a semiconductor. Darker shading represents occupation by electrons

is much smaller than in an insulator, and the conductivity lies between those of conductors and insulators (see Table 6.2).

TABLE 6.2
Typical conductivities and energy differences associated with solid materials

	Conductor	Semiconductor	Insulator
Forbidden energy gap (aJ)	0·004	0·1	1·0
Conductivity $(\Omega^{-1}\,m^{-1})$	10^8	10^2	10^{-8}

6.4.2 Semiconductors

The conduction of electricity through a semiconductor such as silicon or germanium involves electrons moving through the material in one of two ways. The first movement is that of an excited electron, with its energy in the conduction band, which travels freely in the solid. The excitation of such an electron to the conduction band from its position in the valence band leaves behind a vacancy or 'hole' in the valence band, that is in the electronic structure of the atom (point A in Figure 6.14). The second type of movement involves an electron in the valence band (e.g. at point B in Figure 6.14) which acquires enough energy to be able to occupy the hole at A, thereby creating a hole at B. The hole therefore 'moves' from A to B, in the opposite direction to the movement of the electron. Although the movement of charge in this case is due, in fact, to the mobility of the negative electrons, it is convenient to consider the hole as a positive charge, moving in the reverse direction.

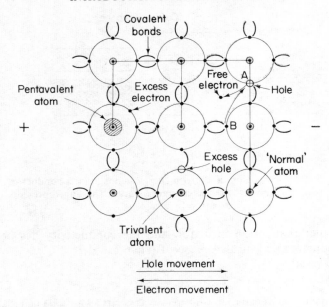

Figure 6.14. Diagrammatic representation of 'holes' and free electrons in a semiconductor, together with the effects on 'hole' and electron concentration of pentavalent and trivalent impurities

We therefore describe conduction in a semiconductor as due to the movement of free electrons and of holes, with the holes moving in the direction of the applied field and free electrons in the opposite direction. The energy required to create a free electron and its associated hole can be provided in several ways. One of these is to supply energy in the form of heat. As the temperature of the semiconductor rises, more holes and free electrons are produced and the conductivity of the material increases. In the *thermistor* (Chapter 2) this effect is used in the measurement of temperature. The energy necessary to create hole–electron pairs can also come from electro–magnetic radiation (e.g. light, X–rays), high–energy particles and strong electric fields. Devices such as solid–state radiation detectors (Chapter 12), photodiodes and photoconductive cells (Chapter 7) work on this principle.

Up to now we have assumed the semiconductor to be a pure element, and in this case the conduction process is termed *intrinsic*. However, the conductivity of a pure semiconductor can be greatly increased by introducing a fraction of a per cent. of impurity atoms, a procedure known as 'doping'. For example, if the additive is a pentavalent atom there will be one electron too many to satisfy the bonding system of the quadrivalent semiconductor (Figure 6.14). The excess electron is easily removed from its atom and becomes a free electron. In this example there are more free electrons than holes in the semiconductor and so the electrons are said to be the 'majority charge–carriers' and the holes the 'minority charge–carriers'. This type of semiconductor, in which the majority charge–carriers are negative, is known as n–type, while the impurity atoms which contribute free electrons are called donors. Examples of donor atoms are arsenic and phosphorus.

In a similar way, if the added impurity is trivalent, excess holes are created (Figure 6.14) so that the majority charge–carriers are positive holes and the semiconductor is p–type. In this case the impurity atoms, such as gallium or indium, are called acceptors.

The addition of minute quantities of acceptor or donor atoms can change the conductivity of the semiconductor by several factors of ten, and radically affect its electrical behaviour.

6.4.3 The p–n Junction Diode

A p–n junction can be formed between p– and n–regions in the same piece of semiconductor (Figure 6.15a). If the junction were made instan–

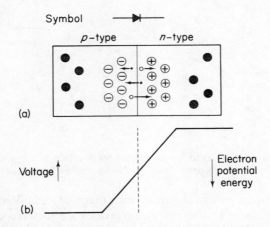

*Figure 6.15.   (a) A p–n junction and (b) the potential gradient set up by the depletion layer. ($\oplus$ and $\ominus$ are positively and negatively charged ions while ● are neutral atoms; • are electrons and o holes.)

taneously, then, because of the concentration differences, excess electrons would diffuse from the n–region into the p–region while excess holes would diffuse in the opposite direction. The region which contains the junction then lacks majority carriers and is called the 'depletion layer'. Previously neutral donor and acceptor atoms in the depletion layer thus become positively and negatively charged, respectively. At regions far enough from the junction to escape its influence, the atoms remain neutral.

The effect of the oppositely charged atoms on either side of the junction is to create a potential gradient (Figure 6.15b) across the junction in a direction which opposes the further movement of majority carriers from either region into the other. A dynamic equilibrium is then set up, in which the effects of diffusion are balanced by those of the electric field.

To make a current flow continuously across the junction it is therefore necessary to reduce the potential barrier by applying an external emf to the semiconductor, in a direction which would make the p–region more positive than the n–region. In this situation the junction is said to be forward–biased (Figure 6.16a), and majority charge–carriers are able to surmount

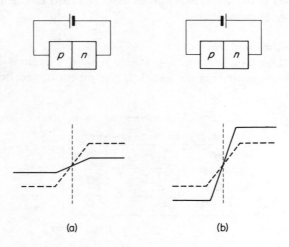

(a)   (b)

Figure 6.16.   (a) Forward bias and (b) reverse bias in a p–n junction diode. The dashed lines are for an unbiased junction

the small remaining potential barrier with thermally acquired energies. An increased emf reduces the barrier and increases the forward current (up to a point where Joule heating in the semiconductor causes irrevocable damage).

If the emf applied to the junction is reversed (Figure 6.16b), the existing potential barrier in the depletion layer is *increased*, and the current which can be carried across the junction by the majority charge–carriers becomes very small. The junction is then said to be reverse–biased. A small reverse current does flow, due in part to minority charge–carriers on each side of the junction, for which the potential difference acts not as a barrier but an inducement to movement. If the reverse bias is increased sufficiently, the free electrons of the reverse current acquire enough kinetic energy to ionize neutral atoms, producing additional free electrons which in turn cause more ionization. A current avalanche thus occurs, at a certain reverse voltage called the breakdown voltage.

The current–voltage curve for a *p–n* junction is shown in Figure 6.17. Between the limits of reverse breakdown voltage and the maximum forward power dissipation, the *p–n* junction only allows a substantial current to pass in one direction, and may therefore be used as a *rectifier* to produce

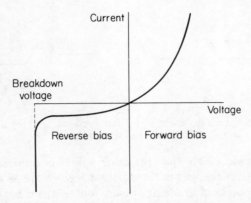

Figure 6.17.   Relationship between voltage and current for a semi–conductor junction diode

direct current from an alternating current supply (see Section 6.5). Advantage is taken of the reproducible breakdown voltage of a *p–n* junction in designing voltage–regulator circuits (such as that shown in Figure 6.28). Diodes designed for this use are called avalanche diodes or Zener diodes, and are at present available for regulating potentials up to 100 V.

6.4.4 *The Junction Transistor*

A transistor consists of a thin layer (c. 1 μm) of *p*– or *n*–type semiconduct–ing material sandwiched between two layers of the opposite kind, thus producing what are called *npn* or *pnp* junctions (Figure 6.18). The inner region is called the *base* while the outer layers are the *collector* and the

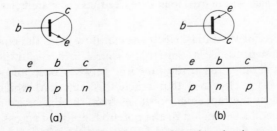

Figure 6.18.   Schematic arrangement of semiconducting material in (a) an *npn* and (b) a *pnp* transistor

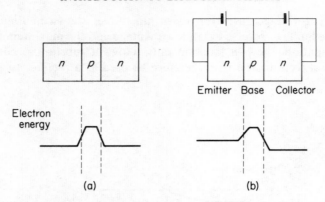

Figure 6.19.   Energy levels in (a) an unbiased and (b) a correctly biased *npn* transistor

emitter. In the case of an *npn* transistor which is unbiased (no voltages applied) the electron energies in the different regions are as shown in Figure 6.19a, since the potential gradient at each junction is a barrier to electrons moving from an *n*–region to the *p*–region. If the junction between emitter and base is now forward–biased, the energy levels are altered (Figure 6.19b) and charge flows as shown in Figure 6.20.

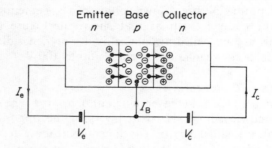

Figure 6.20.   The charge distribution and flow in an *npn* transistor. ⊕ and ⊖ are fixed ions, • are electrons and ○ are holes.)

Because of the forward bias, electrons can flow from the emitter into the base region, constituting an *emitter current*, I_ε. In the thin base region these electrons rapidly reach the junction between base and collector, by diffusion. However, electrons that reach this junction, which is reverse–biased (i.e. against the movement of holes from the base), are rapidly accelerated into the collector region by the potential gradient across the junction, and constitute a *collector current*, I_C. Electrons thus flow from the emitter to the collector, via the base. Some of the electrons in the emitter current recombine with holes before they reach the collector, so that the collector current is less than the emitter current by a certain small amount, which is made up by the *base current*, I_B, which flows directly to the base from the external circuit. I_B is usually of the order of microamperes whereas currents of the order of milliamperes flow in the collector and emitter circuits. It is found that small changes in the base current can produce much larger changes in the collector current, so that the transistor can be used as a current amplifier.

The circuit we have considered (Figures 6.19b and 6.20) is called a 'common–base' arrangement since the collector and emitter circuits have a common connection to the base. More usually the transistor is used in a 'common–emitter' circuit, as shown in Figure 6.21. The current and voltage

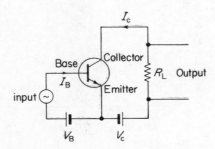

Figure 6.21. A circuit with an *npn* transistor used in the common–emitter configuration. The symbol shown is the conventional representation of a transistor, with the arrow on the emitter indicating the conventional direction of current flow. In a *pnp* transistor the arrow points in the opposite direction and the battery polarities would be reversed

graphs which describe the behaviour of the transistor in such a circuit are called *transistor characteristics*. For instance, the input characteristic (Figure 6.22a) shows how the base current depends on the base : emitter voltage (V_B in Figure 6.21), while the output characteristics (Figure 6.22b) show the variation of collector current with collector : emitter voltage, V_C, for different values of base current. The transfer characteristic (Figure 6.22c) indicates how the collector current varies with base current.

Over a large range of values the transfer characteristic is linear and the ratio I_C/I_B, called the *current gain*, is constant and may be of the order of 1000 or more. An alternating current introduced into the base circuit of Figure 6.21 will therefore cause proportional but amplified oscillations in the collector current. The effective input resistance of the transistor for alternating currents is found by considering the input characteristic. If the base voltage changes by ΔV_B and causes the base current to change by

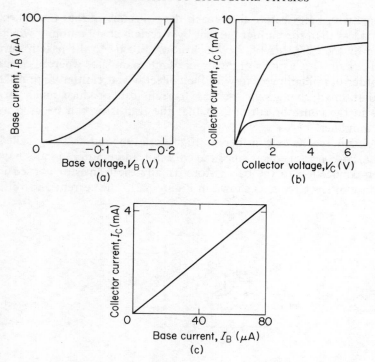

Figure 6.22. Common–emitter transistor characteristics: (a) input characteristic, (b) output characteristics, (c) transfer characteristic

ΔI_B, then the effective resistance to A.C. is $\Delta V_B/\Delta I_B$, that is the *inverse slope* of the characteristic over the range in which the variations occur. In fact, the value of $\Delta V_B/\Delta I_B$ is only a few hundred ohms. On the other hand, the output resistance of the transistor for A.C. can be very high, as shown by the slope of the output characteristic (Figure 6.22b), and may be several megohms. For this reason a high load resistance R_L can be placed in the collector circuit (Figure 6.21) without unduly reducing the A.C. collector current, and the potential difference, $I_C R_L$, which appears across the resistor is an amplified (and inverted) version of the input signal.

Obviously a *pnp* transistor would work in the same way as we have described above, except that the polarity of the applied voltages would be reversed. The emitter in the *pnp* transistor symbol then carries an arrow in the reverse direction to that of Figure 6.21, since conventional current flows *into* the emitter region.

6.4.5 *Other Devices*

Although the junction diode and junction transistor are the best known, there are other useful solid–state devices, of which we will mention only a few. The *unijunction* (Figure 6.23) is a *p–n* junction diode in which the *n* (base)–region has two separate connections, so that a potential difference can be applied between them. The potential at the point in the *n*–region

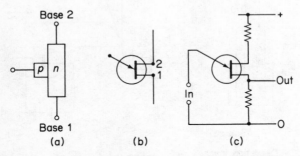

Figure 6.23.   The unijunction transistor: (a) schematic construction, (b) equivalent symbol, (c) connection in a circuit

opposite the *p*–region thus depends on the potentials of base 1 and base 2, and the *p–n* junction only conducts substantially when it becomes forward–biased relative to this point. A *silicon–controlled rectifier* (SCR) or *thyristor* is another type of diode (Figure 6.24) composed of four alternate layers of

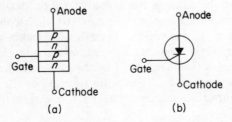

Figure 6.24.   The thyristor or SCR: (a) schematic construction, (b) equivalent symbol

p– and *n*–material, which conducts very little until the applied forward (anode to cathode) voltage exceeds a certain 'breakover' voltage. The diode then switches into a highly conducting state, in which it will remain unless the forward current falls below a certain 'holding' value. A positive voltage applied to the second *p*–region has the effect of reducing the breakover voltage at which conduction begins. A positive pulse applied to this 'gate' can therefore be used to *switch* the diode into its conducting state when the forward voltage is just less than the normal breakover value. The SCR can be regarded as a voltage–operated switch.

An important modification of the junction transistor is the *field–effect transistor* (FET), in which the base region, instead of forming a continuous barrier between emitter and collector, is penetrated by a 'channel' of the emitter/collector material. In Figure 6.25 the channel is of *n*–type material

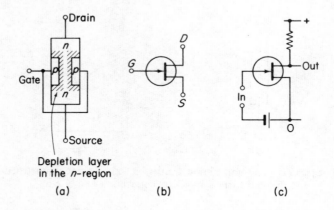

(a)   (b)   (c)

Figure 6.25.   The field–effect transistor. An *n*–channel FET is shown here: (a) schematic construction, (b) equivalent symbol, (c) in circuit

and the transistor is called an '*n*–channel FET'. Here electrons can normally pass freely through the channel from 'source' (*S*) to 'drain' (*D*). If a reverse bias is applied to the base material or gate (*G*), the depletion layer at the boundary between *p*– and *n*–regions increases in thickness, thus effectively reducing the bore of the channel through which the electrons can pass. The electron current is thus controlled by the gate *potential*, so that the gate behaves like the control grid of a vacuum tube. Since the gate is used with reverse bias, the input resistance of the FET is extremely high (hundreds of megohms).

Apart from the diversification in the functions of solid–state devices, another development has been in the direction of so–called 'microminiaturization', to fulfil the demands of the computer industry and space research. To reduce the size of circuits still further the *integrated circuit* (I.C.) has been produced, in which a complete circuit is formed on a small chip of semi-conductor a few millimetres across. This is done by etching away the semi-conductor in selected regions and coating it in others with conducting or dielectric material, to form junctions, resistors, capacitors, and so on. The solid chip is known as a 'monolithic' I.C., whereas 'thin film' I.C.s are entirely formed by evaporating thin films of material on an inert support. In this way, it is possible to produce entire amplifiers which occupy only a few cubic millimetres.

6.5 Circuits and Applications

Before describing a few examples from the myriad applications of electronics we should look first at the circuits themselves, that is the methods of connecting electrically one component to another. When vacuum tubes were predominant, these were mounted in holders on a solid metal chassis and connexions to other components on the chassis were made with copper wire. However, transistor circuits, using low voltages, have generally smaller (and lighter) components which can be mounted on thin insulating wafers or *circuit boards* which are perforated with a rectangular network of holes at a standard 2–mm spacing. The tags or wires of the components can be anchored through the holes and then soldered into the circuit, connexions again being made with wire. A more permanent method of connexion, suitable for mass–production, is the *printed circuit board*. Here a plain plastic board coated with a copper film is treated either photographically or by painting with a special solution, so that on immersion in an etching medium the copper is removed except in lines or areas that will then form the conducting connexions instead of wire. The board is thus 'printed' with the circuit, and the components need only be soldered into place.

6.5.1 *Power Supplies, Filters, Regulation*

Unless it is battery–powered, a circuit usually requires a D.C. supply of power which is obtained from the A.C. mains supply. The conversion of A.C. to D.C., called rectification, is performed by diodes, either of the

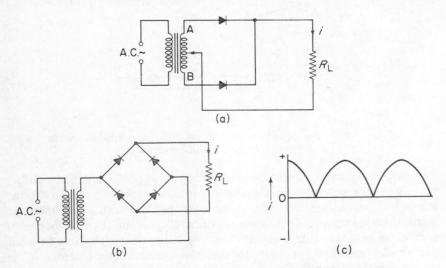

Figure 6.26. Full–wave rectification (a) with a centre–tapped transformer and (b) with a full–wave diode bridge. The output through R_L is shown in (c)

vacuum or p–n junction type. Two typical *full–wave* rectifying circuits are shown in Figure 6.26, both of which transfer power over the complete sine wave of the A.C. current. In (a) a centre–tapped transformer becomes positive alternately at terminal A and at terminal B, relative to the centre point. Thus, first the diode connected to A, then that at B, is forward–biased and conducts while the other diode is reverse–biased and does not. On both half–cycles of the A.C. wave, current therefore flows in *one direction* through the load, R_L. In (b) a transformer without centre–tap is used with a full–wave bridge comprising four diodes, which again ensure a current in one direction through R_L. In both circuits the output voltage across R_L (Figure 6.26c) is unidirectional, but is not a steady value. To get rid of the 'ripple' of the output waveform and produce a smooth D.C. voltage across the load, a capacitor may be connected in parallel with R_L (Figure 6.27a).

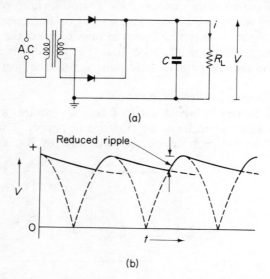

Figure 6.27.   The use of a capacitor in smoothing a rectified waveform :
(a) the circuit, (b) the voltage across R_L

The capacitor charges up to the maximum value of the rectified voltage, but when this voltage falls the capacitor slowly loses its stored charge through R_L, thus maintaining a current through the load even when the rectified voltage is zero. If the time constant, CR_L, of the circuit (Chapter 5) is great enough, the voltage across C and R_L drops little in the time between one ripple peak and the next, so that the voltage output across R_L is 'smoothed', as shown in Figure 6.27(b).

Another way of considering the smoothing action is to say that the rectified voltage waveform (Figure 6.26c) is equivalent to a steady D.C.

value with an A.C. 'ripple' voltage superimposed on it. Since the capacitor has a low impedance for high frequency currents, it therefore acts as a short–circuit for the A.C. ripple current and prevents it passing through R_L. The capacitor across R_L is a crude example of a *low–pass filter*, i.e., a circuit which 'filters out' high frequency current and only allows low frequency or D.C. to pass through the load. Complex networks of capacitors and inductors can be made to act as low–pass, high–pass and band–pass filters, with exact frequency limits between which the transmission of a signal is allowed. They are used, for instance, in telecommunications.

Even though we have a D.C. supply, it is possible that the voltage may vary, due to mains fluctuations or to variations in the current drawn by the remainder of the circuit. It is therefore convenient to *regulate* the voltage by the use of some form of negative feedback. A simple circuit (Figure 6.28)

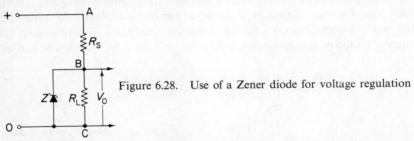

Figure 6.28. Use of a Zener diode for voltage regulation

uses a Zener diode which conducts at a voltage greater than its reverse breakdown voltage, V_{ref}. The normal circuit values are chosen so that if the Zener were absent $V_0 > V_{ref}$. The Zener therefore breaks down and conducts a current, reducing the resistance between B and C until V_0 falls to a value nearly equal to V_{ref}. If V_0 tends to rise, the current drawn by the diode increases and pulls V_0 down, and vice versa. V_0 is thus stabilized at the value V_{ref}, except for conditions in which the Zener no longer conducts at all (e.g. when R_L is short–circuited).

The regulator we have described is a shunt regulator (it acts as a resistance in parallel with the load) but more sophisticated circuits use series regulation, controlling the current in series with the load. Their basic principle is a

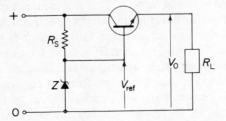

Figure 6.29. Series voltage regulator using an *npn* transistor

good example of negative feedback and is illustrated in Figure 6.29. The voltage V_0 across the load R_L is compared with a steady reference voltage, V_{ref} (provided in this case by R_S and Z). The difference or error voltage $(V_{ref} - V_0)$ is applied to a current regulator in such a direction as to increase the current in R_L and thus increase V_0 until $(V_{ref} - V_0) = 0$. In this circuit the current regulator is an *npn* power transistor in the common–base configuration, and $(V_{ref} - V_0)$ constitutes the base–emitter voltage, which controls the amount of collector current through R_L.

6.5.2 *Amplifiers*

We have already described the action of a single–stage transistor amplifier in a common–emitter circuit (Figure 6.21), and have also encountered the common–base configuration (Figures 6.20 and 6.29). One other possible arrangement for the single transistor remains—the common–collector circuit. (Often the word 'common' in these descriptions is replaced by *grounded*, since the common region is often connected to earth potential.) In the common–collector circuit shown here (Figure 6.30) a *pnp* transistor is used,

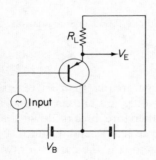

Figure 6.30. A common–collector circuit using a *pnp* transistor

and it can be seen that the *n*(base)–region is reverse–biased, so that the input resistance is high, an advantage in taking signals from a high–impedance source. (The common–collector is often used as the first stage of an amplifier, for this reason.) The output resistance is low, since the emitter circuit is forward–biased.

In this circuit a current will flow in R_L if the emitter potential is greater than the base potential; the effect of the current is to produce such a voltage drop across R_L that the emitter potential falls, until it is practically equal to the base potential. The output (emitter) voltage V_B thus tends to follow the base voltage exactly, and the circuit is often called an emitter–follower. The voltage amplification in this case is therefore just equal to unity, although there is considerable current gain. An alternative way of obtaining a high input resistance would be to replace the *pnp* transistor in Figure 6.30 by a *p*–channel FET.

The features of the three types of transistor circuit are summarized in Table 6.3. It can be seen that only in the common–emitter are both current gain and voltage gain greater than unity, so that it has a high power amplification compa·ed with the other circuits.

TABLE 6.3
Features of the three circuit configurations of a transistor amplifier

	Common connexion		
	Base	Emitter	Collector
Input resistance	low	low	high
Output resistance	high	high	low
Current gain	$\lesssim 1$	> 1	> 1
Voltage gain	> 1	> 1	1
Power gain	> 1	$\geqslant 1$	> 1

When transistors are to be coupled together to form a multistage amplifier, the same problem arises as in vacuum tube circuits—the average output potential from one stage is not convenient as the input (bias) potential for the next. A simple (but inconvenient) way would be to couple the stages through batteries of the required potential, as in Figure 6.31, so that the

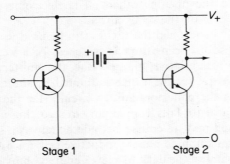

Figure 6.31.   Battery coupling between stages of a transistor amplifier

mean output voltage of stage 1 was reduced to a value just large enough to give the correct base–emitter current in stage 2.

A common method in A.C. amplifiers is to use capacitor coupling (Figure 6.32), in which the base potential of stage 2 is fixed by the voltage divider,

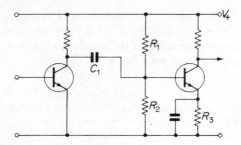

Figure 6.32.   Capacitor coupling between amplifier stages

R_1R_2, and the emitter potential by the capacitor–smoothed voltage across R_3. The coupling capacitor C_1 charges up to the average potential difference between the collector of stage 1 and the base of stage 2 (replacing the battery of the previous figure), and although it prevents direct current flowing, transmits A.C. signals from one stage to the next.

6.5.3 *Control and Switching*

In an experimental situation it is often necessary to control events on the basis of information from a transducer (such as a thermistor or a light-sensitive cell) which supplies information about the environment in the form of an electrical signal. The control often takes the form of switching on or off apparatus such as heaters, refrigerators, lights, and so on, and this is most often done by mechanical switches operated magnetically by the current in a *relay coil*. (The coil may operate several switch contacts at once, closing some and opening others.) The electronic problem is to use the transducer to turn on the relay current.

We have already mentioned that SCRs can be used to switch on a large current. We can also use transistors as switches, for if we increase the base current from zero to a relatively large value, the emitter–collector current increases from zero to a value which is limited by the external resistance and the voltage in the collector–emitter circuit. (I.e. the emitter–collector resistance of the transistor falls from infinity to zero, thus acting as a switch.) The transducer itself may, of course, act adequately as a switch—the resistance of a cadmium sulphide photoconductive cell, for example, may drop from several megohms in the dark to a few hundred ohms when illuminated.

To illustrate methods of switching, we consider the use of a thermistor to control the temperature of, for example, a heated stage on a microscope. In Figure 6.33a, the thermistor, situated near the microscope specimen, is used directly in series with the relay. As the temperature rises, the thermistor resistance drops and the current increases to a point where the relay operates, opening the switch in the heater circuit and turning off the heater. When

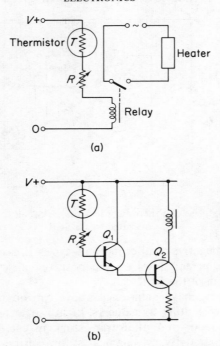

Figure 6.33. Temperature–controlled relay switching (a) directly by a thermistor and (b) after amplification of the thermistor current. The heater circuit is not shown in (b)

the temperature falls, the reverse process occurs, turning on the heater again. Resistor R controls, over a limited range, the temperature at which the relay operates. A more sensitive circuit (Figure 6.33b) uses transistor Q_1 to amplify the thermistor current. The emitter current of Q_1 constitutes the base current of power transistor Q_2, which acts as a switch and turns on the relay when the thermistor current rises.

A better method of temperature–setting uses the thermistor as the fourth resistance of a Wheatstone bridge circuit (Figure 6.34). Two arms of the bridge have the same resistance R, so that when the thermistor resistance R_2 is equal to the setting resistance R_1 no potential difference exists across AB. When the temperature rises and R_2 falls, A becomes positive with respect to B. This potential difference can be applied as shown (or after amplification) to the gate of an SCR, causing it to conduct and thus operate the relay. The bridge circuit is a sensitive detector of resistance changes (Chapter 4) and allows the operating temperature to be set accurately with R_1. An A.C. supply is necessary with the SCR, for with D.C. the SCR, once turned on, would remain on indefinitely. With A.C. the SCR anode

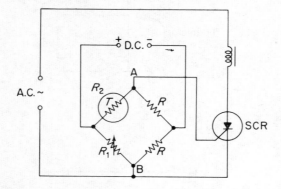

Figure 6.34.   Use of a thermistor in a bridge circuit to control a relay
operated by an SCR

voltage becomes negative on alternate half–cycles, thus continually switching off the current. Current flows on the positive half–cycles if a voltage is applied to the gate of the SCR.

The principles we have mentioned above can obviously be applied, in suitable circuits, with other transducers. Light is detected by photoconducting cells, photodiodes and phototransistors, in each of which the light increases the conductivity. Mechanical strain can be detected in a *strain gauge*, where the resistance of a fine wire is increased when it is stretched. Strain gauges are also used in *pressure transducers*, which detect the small movement of a diaphragm caused by a change of pressure. (Another type of pressure transducer is a ferroelectric crystal (such as barium titanate), in which compression produces a potential difference between the crystal faces.) Methods of temperature–measurement are described in Chapter 2 and the detection of light in Chapter 7.

6.5.4 *Other Uses*

The applications of electronics in oscillating circuits, radio transmitters and receivers and computing logic are beyond the scope of this book, and, probably, outside the direct interest of most biologists. However, we draw attention to two other topics which interested readers may pursue in the references following this chapter. The first of these is the generation and counting of voltage pulses, using monostable and bistable *multivibrator* circuits—a useful technique for counting or synchronizing events, controlling stroboscopes, etc. The second is the use of the *differential amplifier*, a circuit which detects and amplifies the difference between two input signals and is used in situations where a signal has to be distinguished from a 'noisy' background.

Further Reading

I.T.T. Federal Electric Corporation (1966). *Special Purpose Transistors*. Prentice–Hall, New Jersey.

Millman, J., and C.C. Halkias (1967). *Electronic Devices and Circuits*. McGraw–Hill, New York.

Norton, H.N. (1969). *Handbook of Transducers for Electronic Measuring Systems*. Prentice–Hall, New Jersey.

Offner, F.F. (1967). *Electronics for Biologists*. McGraw–Hill, New York.

Zepler, E.E., and S.W. Punnet (1963). *Electronic Devices and Networks*. Blackie, London.

Zucker, M.H. (1969). *Electronic Circuits for the Behavioural and Biomedical Sciences*. Freeman, San Francisco.

PROBLEMS

6.1 A certain photomultiplier has ten dynodes, at each of which, under given conditions, an incident electron causes the emission of three secondary electrons. Assuming perfect efficiency, how many electrons reach the anode for one that leaves the photocathode? If the average secondary multiplication factor is now increased to 3·5, what is the new current gain of the photomultiplier?

6.2 What type of conduction takes place in (a) a pure semiconductor and (b) a 'doped' one? A germanium semiconductor is doped with arsenic atoms; state (c) whether the atoms are donors or acceptors, (d) what type of semiconductor is produced and (e) which are the minority charge–carriers.

6.3 A battery is connected to a p–n junction diode so that the junction is reverse–biased. Which terminal of the battery is connected to the p–side of the diode? Which charge–carriers are free to move across the junction, and are they electrons or holes if they can move from n to p?

6.4 For a common–emitter transistor circuit, the output characteristics of the transistor show that if V_C increases from 2 to 4 V, the collector current increases by 0·1 mA. Calculate the average A.C. output resistance of the transistor over this range. If V_C remained constant, what change instead in the base current would be needed to give the same increase in collector current, if the current gain were 50?

6.5 In a p–channel FET a voltage pulse applied to the gate decreases the source–to–drain current. Say whether the pulse is positive or negative and explain why.

6.6 Suppose you have two junction transistors, one *npn* and one *pnp*, and the necessary other circuit components. Explain how you would connect them in order to amplify efficiently the voltage signal from a high–impedance source.

CHAPTER 7

Light

As we shall see in this chapter, light is only a small part of an enormous range of different electromagnetic radiations which vary in scale from the longest radio waves that originate from distant stars to the highest energy γ–rays emitted by radioactive atoms. However, visible light and its neighbours, infrared and ultraviolet radiation, are biologically the most important radiations of all. Radiation from the sun, for instance, produces a temperature at which metabolic processes can rapidly occur and is also the raw–energy input for photosynthesis, the basis of all food–chains in the living world. Apart from such ecological considerations, light also yields information about our environment—naturally, through the visual process, and artificially, when scientists study the interaction of light with matter.

7.1 The Nature of Light

We know that light is a form of energy, for it can be transformed into heat, it produces chemical changes and it is able under suitable circumstances (the photoelectric effect) to knock electrons out of metals. An interesting thing about light is that its energy is radiant energy, that is energy in transport from a source. Light energy cannot therefore be stored (as light) in the way that one stores, for example, electric charge. Because light is radiated outwards from a source, the light energy is spread over a larger and larger area as the light travels outwards, so that an observer measuring the energy which flows through a given area per second finds that it depends on his distance, d, from the source. If he is far enough away for the source to appear very small (a 'point' source), the detected intensity of the light (energy received per unit area per second) falls off as $1/d^2$, and is therefore said to obey the inverse–square law.

If we look at the shadows cast from objects illuminated by a point source of light, the edges of the shadows appear at first sight to be very sharp, which can be held to prove that light travels in straight lines. In fact, until the seventeenth century, light was thought to consist of streams of particles, much as we think of cathode rays as streams of electrons. However, Huygens showed in 1678 that reflection and refraction of light could be explained by assuming the light to be waves (of something) which spread like the waves begun by a raindrop in a pool of water. In 1802 Young showed in a famous experiment (Chapter 9) that a beam of light can interfere with another in the same kind of way that waves cross over each other in the sea; the experimental results could again be explained by the ideas of Huygens. During the same period, the edges of sharp shadows were found on close examination to consist of light and dark lines (diffraction fringes). These discoveries by Fresnel and Fraunhöfer again supported a wave theory for the nature of light.

From diffraction and interference experiments (see Chapter 9) it is possible to measure the wavelength (λ) of the light used (i.e. the distance between two successive wave crests), and it is found to be different for light which gives different colour sensations to the eye. Since the wavelength is very small a small unit of length, the Ångstrom (Å), is often used in describing it: $1\text{Å} = 10^{-10}$ m $= 0.1$ nm. The wavelength of visible light changes in a fairly smooth way over the range of colours seen, for example, in the rainbow, and varies between about 4000 Å or 400 nm (blue) and 7000 Å or 700 nm (red). To relate these lengths to biological dimensions, we may note that the thickness of a protozoan flagellum (e.g. of *Chlamydomonas*) is about 0.2 μm (i.e. 200 nm).

Although the wavelength of light had been measured some years earlier, it was not until 1864 that Maxwell was able to show what the waves were. By combining all the fundamental laws of electricity and magnetism he showed that it was theoretically possible for oscillations in the strength of both electric and magnetic fields to be transmitted through space as a wave. The theoretical velocity of this wave was found to be equal to the measured velocity, c, of light in a vacuum, which is 2.9979×10^8 m s^{-1}. This and other predictions of the theory, together with the experiments of Herz on the production of radio waves, established that light is an *electromagnetic radiation*. In other words, light (in common with X-rays, γ-rays, radar and radio waves) is a succession in time and in space of oscillations in the magnitude of an electric field and a magnetic field, which are both present at any point in a beam of light. The complete range of electromagnetic radiations is set out in Table 7.1. The wavelength, λ, of the radiation and its frequency, v, are related by the formula $\lambda v = c$.

We have now arrived at the 'classical' understanding of the nature of light, which, accepting electromagnetic wave theory as a complete explanation of known phenomena, was unchallenged until around 1900. However, two obstinate problems which cannot be explained by classical theory led to a change of thought.

The first problem concerned the energy distribution among the wavelengths of light emitted by a hot body. If the intensity of light emitted at different wavelengths is plotted against wavelength for an incandescent

TABLE 7.1
The electromagnetic system

Vacuum wavelength (m)	Other units		Radiation	Origin
10^{-12}	0·01 Å	1·24 MeV	Gamma rays	Nuclear
10^{-11}	0·1 Å		⎰ hard	Inner
10^{-10}	1·0 Å		X–rays ⎱	electron
10^{-9}	10 Å	1 nm	soft	transitions
10^{-8}	100 Å	10 nm		
10^{-7}	1000 Å	100 nm	Vacuum ultraviolet	
	1700 Å	170 nm		Outer
			Ultraviolet	electron
	4000 Å	400 nm ⎱		transitions
			Visible	
	7000 Å	700 nm ⎰		
10^{-6}	10 000 Å	1000 nm		
	10 000 cm^{-1}	1 μm		Molecular vibrations
			Near infrared	
10^{-5}	1000 cm^{-1}	10 μm		Molecular
10^{-4}	100 cm^{-1}	100 μm	Far infrared	rotations
10^{-3}	300 GHz	1 mm		
10^{-2}	30 GHz	1 cm	Microwaves	Electron spins
10^{-1}	3 GHz	10 cm		Nuclear spins
1	300 MHz			
			VHF radio	
10	30 MHz	10 m		
			Short–wave band	
10^2	3 MHz	100 m		
			Medium–wave band	
10^3	300 kHz	1 km		
			Long–wave band	

body radiating with maximum efficiency (a 'black' body—see Chapter 2), the solid curve shown in Figure 7.1 is obtained. This experimental curve, the radiation spectrum of a black body, could not be explained by wave theory, which predicted a spectrum described by the dashed curve. At best this agrees with experiment at long wavelengths (the red end of the spectrum) but at worst it predicts an ever increasing light intensity as the wavelength decreases through the ultraviolet end of the spectrum. This 'ultraviolet catastrophe' showed that wave theory did not have all the answers to optical phenomena. The experimental curve was explained by Planck in 1901, but his explanation involved an idea which was a complete departure from classical theory. To explain the results, he had to assume that the energies of molecules in a heated body are *quantized*. That is, the energies only have values which are multiples of a certain basic amount,

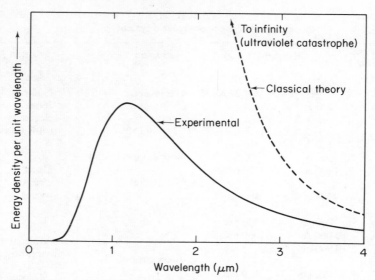

Figure 7.1.   The distribution of energy in the spectrum from a perfect
emitter (black body) at 2500 K

or *quantum*. (To give an analogy, monetary systems are quantized; all
prices are multiples of the smallest coin used in the system and we cannot,
for example, pay someone 0·37 of a penny, although in weight, which is an
unquantized system, we can read 0·37 of a gramme on a balance.)

On Planck's theory, when a molecule or atom loses energy it does so
in steps of one quantum, at the same time changing this quantum of energy
into radiation. The energy (E) in one quantum of radiation is related to
the frequency of oscillation, v, of the radiated light by Planck's law,

$$E = h\,v,\qquad\qquad(7.1)$$

in which the physical constant, h, is called the Planck constant and has
the value $6·626 \times 10^{-34}$ J s. Planck's theory predicted a curve very close
to the experimental one of Figure 7.1 and raised the possibility that light
was not a long train of waves but consisted of separate 'packets' of energy.
Each 'packet' would contain a quantum of energy, the magnitude of which
would depend on the frequency of the light.

The second experimental discovery which demonstrated an inadequacy
of the wave theory was the finding by Lenard, in 1899, that light could
release electrons from the surface of metals. This photoelectric effect has a
so–called threshold frequency, for light with a frequency below this critical
value ejects no electrons no matter how bright the light which falls on the
metal surface. Classical theory cannot account for such a threshold effect,

but Einstein used Planck's quantum theory to explain this and other properties of the photoelectric effect. He proposed that, in addition to molecular
energies, light energy also was quantized in units of magnitude $h\nu$—the
units we now call *photons*. If the frequency of the light is high enough,
a photon has enough energy to release an electron from the forces which
keep it within the metal surface, but below a certain value of ν the energy,
$h\nu$, of a photon is insufficient so that, no matter how many photons per
second arrive in the light beam, none is able to eject an electron. This accounts
for the threshold effect.

We now have two ways of predicting the behaviour of light, depending
on the type of phenomenon we are investigating. We can consider light
to be a radiated system of waves or a stream of energy quanta (photons).
This wave–particle duality, as it is called, does not imply a conflict between
the two approaches; each idea of the nature of light is a model, or approximation, of a more complex reality, and the choice of the model which works
depends on the conditions of a particular experimental situation.

7.2 The Physics of Light Emission
7.2.1 Excitation
A source of light is matter in which the atoms or molecules are said to
be *excited* to a higher energy than that at which they are stable. A stable
atom, for instance, is excited when one of its electrons moves from one
orbit to another in which its total energy (kinetic + potential) is higher.
After this energy change or *transition*, the atom is said to be in an excited

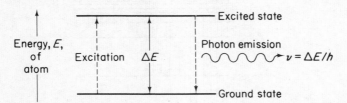

Figure 7.2.   Energy transitions and photon emission from an atom

state (Figure 7.2). If the electron then returns to its original energy level,
the atom as a whole is returned to its stable state of minimum energy (its
'ground' state) and the excess energy, ΔE, which the atom loses in this
transition can appear as a photon of radiated energy. In this case, the energy
of the quantum is already determined by the difference in energy levels
(ΔE) involved in the transition, so that Planck's law can be used in this
case to tell us the frequency of the emitted radiation. Since $\Delta E = h\nu$, the
greater the energy jump, the greater is the frequency of the radiation and,

conversely, the smaller its wavelength λ, since the two are related by the equation $c = \lambda v$. In general, then, large energy transitions correspond to short–wavelength radiation (X–rays, ultraviolet light) and small energy transitions to long–wavelength radiation (infrared light, microwaves).

The methods of exciting matter are various, but obviously they are all methods of communicating energy to atoms or molecules. Heat, for instance, gives rise to high–energy collisions between molecules, while an electric spark or discharge in a gas causes collisions of atoms with ions. Light itself can act as an exciting agent when atoms or molecules absorb photons (e.g. in fluorescence, phosphorescence and the operation of a laser), while in chemiluminescence chemical energy is converted directly into the radiation from excited molecules.

7.2.2 *Emission*

We shall now survey the different types of energy transition in atoms and molecules and describe in each case the kind of spectrum that arises and its constituent wavelengths.

To begin with, let us consider electron transitions in free atoms (i.e. atoms in a gas). In each atom the electrons surrounding the nucleus can only have certain energies, since the energy levels are quantized as shown in Figure 7.3. When the atom is excited, an electron moves to a higher energy level, returns to a lower one (not necessarily the one it vacated, if another is vacant) and a photon is radiated of frequency corresponding to the transition energy. In general, as shown in Figure 7.3, it is transitions between

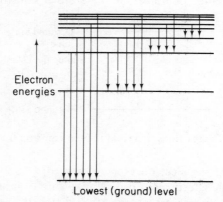

Electron energies

Lowest (ground) level

Figure 7.3.   The quantized energy levels of a simple atom, showing some of the possible radiative transitions

the outer orbitals of the atom which are of the lowest energy and correspond to frequencies in the ultraviolet and visible regions of the spectrum (Table

7.1). Transitions involving the innermost electrons in atoms of high proton number are of much larger energy and the radiation involved then is X–radiation.

When the light radiated from free atoms is split up by a prism into its various frequencies to give a spectrum, we see a *line spectrum* (Figure 7.4).

Figure 7.4. A line emission spectrum from an iron arc discharge

This follows directly from the quantization of the electron energy levels, since a transition from one level to another must give a single radiated frequency which, after the passage of the light through a prism, gives a separate line image of the entrance slit of the spectrometer (Section 7.5.3). It can be seen from Figure 7.3 that the energy levels crowd together for the outer, high–energy, electron orbitals, which means that the number of possible transitions with high energy differences is greater than the number of possible small energy transitions. The spectrum reflects this situation, for the spectral lines crowd together at the high–frequency (blue) end.

When atoms are combined together in molecules, the formation of molecular orbitals, in which the electrons are shared between atoms, gives rise to more allowed energy levels and a consequent increase in the possible number of radiative transitions. The effect of this and of the molecular movements that are discussed below is that the spectrum from a molecular gas is a multiplicity of spectrum lines which group together and appear as bands of light, giving a *band spectrum* (Figure 7.5).

Figure 7.5. A band emission spectrum from nitrogen

An important consequence of the joining together of atoms is that extra forms of energy can appear which depend on the existence of the molecule itself. These are the energy of molecular vibration and the energy of molecular rotation. For a particular molecule these energies only have certain allowed values, so they can be represented by an energy–level diagram for the molecule similar to that for the electrons in an atom. When the molecule makes a transition from one energy to another, it emits or absorbs one quantum of radiation.

In the case of molecular rotations, we can regard the molecule as a rigid body which in a gas is free to rotate with various kinetic energies, E_{rot}, determined by the mass and size of the molecule. A schematic diagram of the rotational energy levels is given in Figure 7.6. Since the frequency

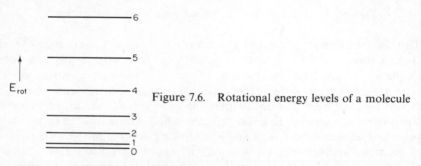

Figure 7.6.   Rotational energy levels of a molecule

of radiation emitted or absorbed when the molecule makes a transition is proportional to the change in energy, it is possible by studying the radiation to find the moment of inertia (Chapter 1) of the molecule. The energy changes involved here are fairly small compared with electron transitions, so the frequency of the radiation is smaller and its wavelength longer than that produced by radiating atoms. For small, light, molecules the rotational transitions correspond to wavelengths in the range from 50 μm to 350 μm. As shown in Table 7.1, this covers the electromagnetic spectrum in what is called the 'far infrared' region, and borders on the microwave region.

Since we have begun to consider infrared radiation, we should mention here another unit which, like the Angstrom, has been' used for convenience by spectroscopists but is principally applied in the infrared region. This is a unit of reciprocal wavelength, which is derived by dividing one by the wavelength of the radiation. Traditionally, spectroscopists have expressed the wavelength in centimetres before performing this division, and the quotient obtained is called the *wavenumber* ($\bar{v}$) of the radiation. Thus

$$\bar{v} = 1/\lambda' = 1/(100\ \lambda), \tag{7.2}$$

where λ' is the wavelength in cm or λ the wavelength in m. On this definition $\bar{v}$ is numerically equal to the number of wavelengths in 1 cm of the light path. (Since the frequency v of the radiation is given by c/λ we can see from equation (7.2) that the wavenumber is proportional to frequency.)

As an example, the range of radiation we quoted for rotational transitions, with wavelengths of from 50 μm to 350 μm, would have wavenumbers of from 200 cm^{-1} to 29 cm^{-1}.

In addition to rotating a molecule is also capable of vibrating, as it is not, in fact, a rigid body but an assembly of atoms held together by bonds of varying elasticity and flexibility. The atoms thus have a certain freedom of relative movement, and the possible number of independent types of movement (or degrees of freedom) of the atoms within a molecule increases with the number of atoms present. It can be shown that a molecule with N atoms has $3N - 5$ or $3N - 6$ degrees of freedom, depending on the molecular shape. For instance, a diatomic molecule ($N = 2$) can perform only one type of vibration, in which the atoms move towards or away from each other along the line joining their centres; this is an example of a so-called 'stretching' vibration.

TABLE 7.2
Infrared absorption peaks for various atomic groupings

Group	Vibration	Approximate wavenumber of absorption peak (cm^{-1})
$\geq$ C$-$H	C$-$H stretch	2900
$\geq$ C$-$H$\updownarrow$	C$-$H deformation	1375
$>$ C $\diagup^{H}\diagdown_{H}$	CH$_2$ scissor deformation	1460
$>$ C $\diagup^{H\downarrow}\diagdown_{H\downarrow}$	CH$_2$ rock deformation	720
$>$C $=$ C$<$	C $=$ C stretch	1620–1680
$=$ C $=$ O	C $=$ O stretch	1680–1780
$-$O$-$H	O$-$H stretch	3600
$>$N$-$H	N$-$H stretch	3310
$>$N$-$H$\updownarrow$	N$-$H deformation	1550

(*Note* : the arrows indicate the direction of the vibration.)

Each degree of freedom has its own vibration frequency, which depends on the strengths of the bonds involved and the masses of the vibrating groups of atoms. Because the vibrations involve the movement and separation of electric charges, each vibration corresponds to an electric dipole of oscillating strength which is capable of emitting or absorbing electromagnetic radiation (Chapter 5). The frequency of the radiation is equal to that of the vibration and typical values lie in the range corresponding to wavenumbers of from 100 cm^{-1} to 4000 cm^{-1}, or wavelengths from 100 μm to 2·5 μm, which comprise the 'near infrared' region of the spectrum. (Values of the wavenumber for the vibrations of various organic groups are given in Table 7.2.)

In summary, then, an atomic gas only gives rise to electron transitions, which correspond to line spectra in the visible and ultraviolet region, while a molecular gas is capable, in addition, of transitions which correspond to infrared wavelengths and, in general, gives a banded spectrum. In a hot liquid or solid the interaction between one molecule and another is much greater than in a gas, with the result that the otherwise clearly defined energy levels of atoms and molecules become much more uncertain in value. In this situation, the energy difference in a transition can have a range of values, and the corresponding radiation has a range of wavelengths instead of the single one that would give a line in the spectrum. The overlapping wavelength ranges due to different transitions produce, therefore, a *continuous spectrum*, which in the visible region appears as a continuous band of light, of colours which smoothly change from blue through to red. The intensity variation in the continuous spectrum depends on temperature, and is shown for a 'black' body in Figures 7.1 and 2.12.

7.2.3 *Fluorescence*

When a molecule is excited, its energy is increased to one of the quantized

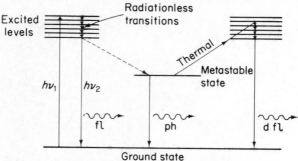

Figure 7.7.   Energy transitions during fluorescence and phosphorescence, where fl is fluorescence, ph is phosphorescence or stimulated emission, and d fl is delayed fluorescence. Note that frequency v_2 of fluorescence is less than v_1 of exciting radiation

energy levels which lie at a higher energy than the normal, or 'ground', state of the system. When the molecule is in a solid or liquid, the interaction of one atom with another results in the excited energy levels being many and close together, so that it is possible for the excited molecule to lose energy in continual small energy transitions from one level to the next, until it reaches the lowest excited state (Figure 7.7). Because the energies involved are small, they can be transferred from one molecule to another by vibration or collision, so that the excited molecule is able to lose some of its energy in this way, rather than by emitting a photon; it is said to make radiationless transitions.

When the molecule reaches its lowest 'allowed' excited state, it may then lose its remaining excess energy by emitting a photon and returning to the ground state. (The 'life–time' of the molecule in its excited state may be as small as 10^{-8} s.) When the original excitation is produced by radiation, it is obvious from Figure 7.7 that the energy and therefore the frequency of the emitted photon is always less than that of the exciting photon. The emission of radiation in the way described is called fluorescence. One application of the principle is in fluorescence microscopy, where the specimen, stained with a fluorescent dye, is illuminated with ultraviolet light: the dye fluoresces at a lower frequency and longer wavelength than the exciting radiation, and the emitted radiation is visible to the eye.

Another way in which the molecule may lose its remaining energy is by transition to a lower energy level, from which further transitions to the ground state are 'forbidden' by the rules of quantum theory. What this means is that there is normally only a very small probability of the molecule making a transition from this lower level directly to the ground state, so the molecule is almost stable at this level, even though it possesses excess energy. Such a situation is called a metastable state.

Molecules in a metastable state may eventually reach the ground state again in several ways. They may, after a time, gain thermal energy and return to a higher energy level from which they can make a direct transition— giving *delayed fluorescence*. They may make the 'forbidden' transition spontaneously, after an average life–time in the metastable state of 10^{-3} s or even longer, depending on how small the probability of such a transition may be. In this case, the emitted photons are of smaller energy and longer wavelength than in fluorescence; the phenomenon of this delayed emission is called *phosphorescence*. Alternatively, the molecules may be induced to make direct transitions from the metastable state to the ground state by the incidence of light of the same frequency as that given by the transition concerned; this is known as *stimulated emission*.

The light source which uses the phenomenon of stimulated emission is the *laser* (Light Amplification by Stimulated Emission of Radiation). The principles of the system can be illustrated by describing the pulsed ruby laser. In one form (Figure 7.8) a ruby rod about 10 mm in diameter by 100 mm in length is surrounded by a helical xenon flash tube. The ends of the ruby rod are made optically flat and parallel, and while one end is made completely reflecting with a coating of silver, the other is only partially

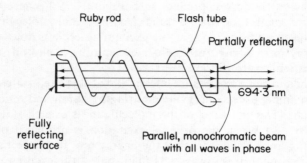

Figure 7.8.   Schematic diagram of laser construction

reflecting. A light pulse from the flash tube lasting, say, 500 μs is sufficient to excite molecules in the ruby rod to a band of energy levels (Figure 7.9) from which they make spontaneous transitions preferentially to a metastable level, so that the majority of excited molecules end up in this state after an

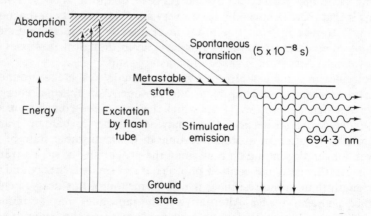

Figure 7.9.   Energy levels and emission of light in a ruby laser

average life–time in the higher band of 5×10^{-8} s. Now the normal life–time of the molecules in the metastable state is fairly long (3×10^{-3} s), so it is possible, if the exciting flash is intense enough, to have more molecules per unit time getting into this state than are escaping from it. The result (an essential one for the laser operation) is that we obtain more molecules in the metastable level than in the ground level, which is the reverse of the usual situation and is called *population inversion*.

When a molecule does make a transition from the metastable state to the ground state, it emits a photon with a wavelength of 694·3 nm. Now such a photon has, of course, exactly the right energy to raise a ground-state molecule back to the metastable level (in which case the photon would be reabsorbed in the ruby rod). But because of the population inversion, the number of molecules in the ground state is reduced and with it the possibility of the photon's absorption. In fact, the photon is now much more likely to interact with the increased number of molecules in the metastable level and stimulate them to return to the ground state, themselves emitting photons of the same wavelength. The result of this stimulated emission is an 'avalanche' effect: one photon induces the emission of another photon, these two photons in turn stimulate emission of another two, those four produce another four, and so on. The light produced, reflected up and down the ruby rod, rapidly induces all the molecules to return to the ground state and the energy is released in a single flash of red light, all at the single wavelength of 694·3 nm, which escapes through the partially reflecting end of the rod.

The characteristics of the laser which make it distinctive are the following. First, energy over a wide range of wavelengths from the flash tube is concentrated into energy released at a single precise wavelength (monochromatic) from the laser. Secondly, because the energy of a pulsed laser is released in a very short time, the intensity during the pulse is very high, of the order of 10 kW mm^{-2}. Thirdly, because of the construction of the laser system, the light issues in a very narrow, accurately parallel pencil of rays, so that the energy is concentrated in a very small beam width even at large distances from the source. Finally, an important characteristic of the stimulated emission is that it is all *coherent*, that is all the light waves from the source issue in step with each other (see Chapter 9), which has important consequences for the laser's use in optical systems.

In medicine the concentrated energy of the laser has been employed therapeutically, in 'welding' detached retinas back to the eyeball; this is done by using the eye lens itself to focus the laser beam on selected points of the retina, where the absorbed energy is released as heat. The laser has also been used in research to damage microorganisms in selected small areas under the microscope. It is possible to direct the laser beam 'backwards' through the microscope so that the energy is focused into a diameter of a few μm, which is small enough, for instance, to damage cellular components, such as mitochondria or selected parts of the flagellum of a protozoon. This technique has been used to investigate the function of the damaged structures. The coherent quality of laser light has been used in optical systems for the analysis of electron–microscope pictures of viruses.

7.3 Some Biological Aspects

Since sunlight is the prime factor in the ecology of this planet, we shall begin this section by considering the sun and its radiation. The diameter of the sun's disc subtends about half a degree of arc as viewed from the earth, and the distribution of energy in the solar spectrum is fairly close to that given by a black body at approximately 6000 K. At a distance in space equal to the mean distance between earth and sun, the total solar radiation (i.e. including all wavelengths of the spectrum) is 1·4 kW per square metre of receiving surface normal to the sun's rays. Over the area which the earth presents to the sun this amounts to a total power of about $1·8 \times 10^{11}$ MW incident in the upper atmosphere of the planet, about equal to the output of two hundred million large power stations.

Of course, not all this energy from the sun is retained (some of it is reflected or re–radiated), and certainly the wavelength distribution in the energy which reaches the earth's surface is quite different from that in the original solar spectrum. The constituents of the atmosphere, in particular water,

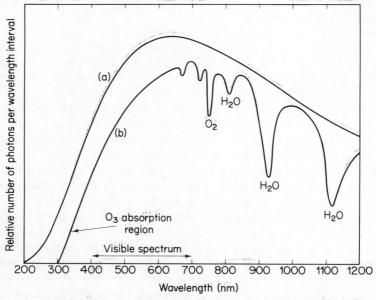

Figure 7.10. The effect of atmospheric constituents on sunlight reaching the earth. Curve (a) is light incident on the upper atmosphere, while curve (b) is light at sea–level (low humidity, good visibility). (Adapted with kind permission from H.H. Seliger and W.D. McElroy, *Light : Physical and Biological Action*, Academic Press, 1965)

carbon dioxide, oxygen and ozone, each absorb energy in different ranges of the spectrum, so that the solar energy reaching the earth's surface at sea-level is severely reduced at particular wavelengths, as shown in Figure 7.10. Fortunately for the present life-forms on earth, the ozone in the upper atmosphere absorbs nearly all the harmful ultraviolet radiation below wavelengths of 300 nm. It is interesting that the peak absorption of ozone occurs at a wavelength very close to the value of 260 nm at which nucleic acids absorb ultraviolet energy and suffer mutation. Some scientists surmise that the appearance of ozone in the atmosphere protected the genetic material of the earliest life on earth from the effects of continuous mutation, and thus allowed the exact replication of organisms to begin.

While ozone, oxygen and nitrogen cut off the radiation at the ultraviolet end of the spectrum, the effect of water vapour and carbon dioxide in the atmosphere is to absorb at the other end, in the near infrared. Water vapour has several absorption wavelengths, from 900 nm (just outside the visible range) upwards, while carbon dioxide begins to absorb from about 2000 nm. The overall effect of the atmosphere, then, is to prevent solar radiation reaching the earth's surface directly, except within a narrow 'window' of the spectrum between 300 and 1000 nm. It is obviously not a coincidence that the range of radiation that our eyes can see (from 400 nm to 700 nm) falls just within this window.

Another point of interest is that within the range of solar radiation which is received at sea-level the intensity (estimated in terms of photons per square metre per second) is a maximum at wavelengths in the centre of the range, around 650 nm, and falls off fairly symmetrically on either side. Therefore in photosynthesis, in which a green plant needs about nine photons of visible light to convert a carbon dioxide molecule into carbohydrate, we might expect the plant to use a pigment which absorbed the maximum number of available photons. In fact, most chlorophylls have one of their absorption peaks at around 650 nm to 700 nm.

The photosynthetic process is a very complex one, both chemically and in its response to light of different wavelengths. At least two pigments are involved, one of which (usually chlorophyll a_1) absorbs photons at a far-red wavelength of about 700 nm while the second pigment absorbs at a shorter red wavelength. For efficient photosynthesis photons must be absorbed by both pigments. It is believed that chlorophyll a_1, oxidized on excitation, is concerned in the production of the energy-rich molecule adenosine triphosphate (ATP), which is used in the conversion of carbon dioxide to carbohydrate. The second pigment is ultimately responsible for the release of oxygen which accompanies the carbon dioxide fixation. In addition, accessory pigments (chlorophyll b in land plants, other pigments in red and blue-green algae) exist which absorb light energy and are able to pass it on to the photosynthetic pigments. It is the accessory pigments which enable the plant to adapt to its particular light environment. For instance, about 10 m below the surface of the sea the light is predominantly green, but the red pigment of a red alga absorbs this green light and enables the plant to continue photosynthesis where green algae are unable to thrive.

The converse of photosynthesis, where light energy is transformed into chemical energy, is chemiluminescence, in which the energy transformation is in the opposite direction. In the latter phenomenon a chemical process leads to the emission of radiation which has a higher intensity (over a certain wavelength range) than a black body would give when at the temperature of the reactants. The light emitted is thus due not to thermal excitation of the molecules but to chemical excitation.

The type of luminescence that occurs in living organisms is called bioluminescence, and is invariably an enzyme–catalysed form of chemiluminescence. Visibly luminous forms occur among fungi, bacteria, jellyfish, worms, crustacea, fish, and so on, but the best–known example is that of the fire–fly, in which luminous flashes are used as a device for recognition in mating. The substances responsible for light emission have been extracted and purified, so that the reaction can be studied in the laboratory. The enzyme is called luciferase and the substrate on which it acts is an organic molecule of low molecular mass, called luciferin. In the presence of magnesium ions and ATP the enzyme and substrate join together to form a luciferin complex, which reacts with oxygen and is raised to an excited state by the energy released in the oxidation. The complex returns to a stable state with the emission of light with a peak intensity at 562 nm (yellow–green). The complete process specifically requires the presence of ATP and the light output is directly proportional, for small concentrations, to the amount of ATP that is used. The luciferin reaction can therefore be used in the laboratory as a means of detecting and measuring minute concentrations of this important molecule. By using a sensitive light detector, concentrations as low as 10^{-6} mol m^{-3} can be measured with accuracy and amounts of ATP of the order of 10^{-13} mol can be detected.

7.4 The Detection and Measurement of Light

The detection of light essentially involves a process of energy conversion, in which the photons incident on the detector are absorbed and a proportion of the energy changed into a more useful form. For instance, if light falls on the blackened surface of a thermometer, most of the light is absorbed and converted into heat, which is registered by a rise in temperature of the thermometer. This type of detection is called thermal detection, and has the advantage that if the detector surface is 'perfectly black' (i.e. it absorbs all the incident radiation at all wavelengths) the thermal detector gives the same response to a given energy from any part of the spectrum. For this reason the thermal detector is called a *non–selective* detector.

Most other detectors are selective in character, as the efficiencies of other processes of energy conversion depend on the wavelength of the radiation. We can roughly divide these processes into optical, electrical and chemical ones. Optical processes include, for example, fluorescence, where ultraviolet light, invisible or injurious to the eye, is converted into light of visible wavelength. In electrical processes the photon energy is used to eject an electron from a surface (photoelectric effect), raise it to a higher energy level (photoconduction) or transfer it across the junction of two materials (photovoltaic effect). As chemical processes we have photography, the visual process in the retina of the eye and chemical 'actinometers', in which light levels are measured by the chemical changes produced in certain inorganic solutions.

7.4.1 Noise

It is difficult to compare the sensitivities of different detectors because each may operate in a different region of the spectrum or produce a different kind of output (current, voltage, photographic density, moles of released ions, etc.). One figure which, in principle, can be calculated for any detector is the 'noise–equivalent power'. Random fluctuations in response ('noise') obviously limit the minimum response or signal which can be registered efficiently by a detector, for if the response of the detector to the incident radiation is of the same order of magnitude as the random fluctuations (i.e. a 'signal–to–noise' ratio of one) it will be difficult to distinguish the real response from the background noise.

The noise–equivalent power (N.E.P.) of a detector is the power of the incident radiation ($J s^{-1}$ or W) which would produce a signal just equal to the averaged noise level, and thus gives an idea of the minimum power which the system can detect. Values of the N.E.P. are around 10^{-11} W for thermal detectors, 10^{-9} W to 10^{-14} W for photoconductive devices and 10^{-16} W in photomultipliers (photoelectric devices). It is interesting that a power of about 10^{-16} W is the minimum which must be accepted by the pupil of the eye to enable a white point source (such as a star) to be seen in extrafoveal vision.

7.4.2 Modulated ('Chopped') Radiation

Thermal detectors in particular, and other detectors to some degree, suffer from two disadvantages: there is probably some 'standing' response, even in the absence of radiation (zero error), and also this response may drift slowly with time, as external conditions change (zero drift). One method of overcoming this is to build compensating systems in which two detectors are used in opposition, so that external effects on both detectors tend to cancel out while only one detector receives the radiation. Another technique, known as 'chopping', is to modulate or change the intensity of the radiation that falls on a single detector, so that it receives a fluctuating impulse rather than a steady one. One simple form of 'chopper' is a continuously rotating disk placed in front of the detector. The disk has regularly spaced sectors cut out of it, which enable the detector to receive the radiation intermittently. The output of the detector is therefore a steady signal, due to external and other factors, plus a *fluctuating* signal when radiation is present. By electrical methods (such as using a transformer) the fluctuating signal can be isolated from the steady one and separately amplified. This removes 'zero error' and also 'zero drift', if the drift is very slow compared with the frequency of chopping. It also helps to reduce the effects of 'noise' in the detector.

7.4.3 *Thermal Detectors*

Thermal devices are used in the infrared region of the spectrum, as the efficiency of other, selective, devices decreases as one moves further into this region and away from the visible one. As we have seen, the thermal detector is a type of thermometer which registers the rise in temperature of a detecting surface. The surface is coated by evaporation with finely divided metal (such as gold) which is 'black' from the ultraviolet region right up to wavelengths of 40 μm (250 cm^{-1}).

In the *bolometer* the detecting surface is a thin strip of platinum foil of which the electrical resistance increases with temperature. To compensate for temperature increases which are not due to the radiation, two strips are often used, connected as two adjacent arms of a Wheatstone bridge circuit. Only one of the strips receives the radiation, while both are affected by external temperature changes, the effects of which cancel out since they affect the resistance in both arms of the bridge. Radiation falling on one platinum strip produces an increase in that arm and consequently a current in the bridge galvanometer (Chapter 4). The bolometric strips are enclosed in a vacuum to reduce external effects. Instead of platinum, more modern bolometers use thermistor materials which have a much greater (negative) temperature coefficient of resistance.

In the *thermocouple* a rise in temperature of the detecting surface is communicated to the 'hot' junction of two metals or semiconductors in a thermoelectric circuit, while the 'cold' junction is at room temperature and shielded from the incident radiation. The difference in temperature of the junctions produces an emf in the circuit (Chapter 2) which is a measure of the power of the radiation received. To increase the response of the system, several thermocouples are usually arranged to act in series, so that the emfs add together. This arrangement is called a *thermopile*, an example of which is shown in Figure 7.11. Here the 'cold' junctions of each pair of metals are cemented to a support which does not receive the radiation, while the 'hot' junctions are cemented to the blackened detecting surface. (Electrically insulating cement is used to prevent short–circuits.) The sum

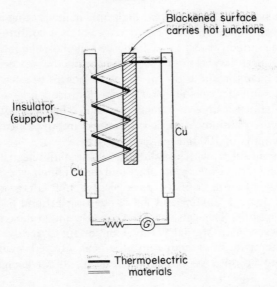

Figure 7.11.   Construction of a thermopile with a 'straight line' detector surface

of the emfs is detected by a galvanometer circuit. To reduce external temperature influences the thermopile is enclosed in a vacuum tube, which, however, must have a suitable window to allow the radiation to enter.

Another, completely different, type of thermal detector is the *Golay cell*, in which the temperature measurement is effected by what is, in essence, a gas thermometer. A small chamber with dimensions of a few mm contains air or a gas and receives the radiation through a suitable window. Within the chamber is an aluminized detector surface which, on absorbing the radiation, heats the gas, which in turn tends to expand. Another part of the chamber is closed by a thin, flexible, metallized membrane which acts as a mirror, so that the expansion of the gas distorts the mirror and disturbs a complex optical system of which it is part. The optical changes are converted into an electrical signal.

7.4.4 *Electrical Detectors*

Detectors which depend on photon–electron interactions are inherently selective since they depend on the energy of a single photon, which varies with the radiation frequency. For this reason the photomultiplier, the photoconductive and photovoltaic cells tend to be used for ultraviolet and visible light rather than the infrared, where photon energies are low. The sensitivity of the detector decreases to zero when the photon energy drops below a 'threshold' value. Since the principles of these devices have been discussed elsewhere (Chapter 6), we shall only mention a few of their optical properties in this chapter.

The photoconductive cell is, in fact, used in the near infrared region as well as at shorter wavelengths, as semiconductor materials have been found which have threshold or 'cut–off' wavelengths of a few μm. Materials such as lead sulphide, selenide or telluride are used, and are often cooled to liquid air temperatures to extend their cut–off wavelengths farther into the infrared. Germanium 'doped' with added impurities can also be used up to wavelengths of 100 μm in a similar way.

The photomultiplier, which makes use of the photoelectric effect, is ideal for ultraviolet and visible radiation, and has a high sensitivity in these regions. Most photomultipliers have a response curve which shows a maximum at around 400 nm and falls to zero at around 650 nm, the limit of the visible region. The sensitivity is thus very much dependent on wavelength. However, with special cathode materials to receive the radiation, the cut–off can again be extended into the near infrared region. (The silver–caesium photocathode, Ag–O–Cs, will respond to wavelengths up to 1 or 2 μm.)

Photovoltaic cells are used predominantly in the visible region of the spectrum—in portable apparatus where it is inconvenient to provide the supply to run a photoconductive cell. For instance, they are used in some types of photographic exposure meter.

7.4.5 Chemical Detectors

Light–induced chemical changes occur both in the retina of the eye and in the sensitive emulsion on a photographic film. The wavelength range within which these detectors can work is limited, particularly in the infrared. The eye responds to radiation between about 400 nm and 700 nm, with a peak response about midway between. (The cut–off at the violet end of the spectrum is due principally to the absorption of ultraviolet light by the lens of the eye, for the retina itself is sensitive to much shorter wavelengths.) The photographic emulsion, on the other hand, is very sensitive to ultraviolet light, and even to X–rays, and by adding dyes to the emulsion it is made sensitive throughout the visible spectrum. Special infrared–sensitive emulsions can be obtained which record images with wavelengths up to 950 nm to 1200 nm. Although photographic emulsions are less sensitive to light than a photomultiplier, they have the advantage, for some uses, that they are *integrating* devices, for which the response depends on the total radiation received within the exposure time. In addition, of course, they provide a permanent record of their response.

Another integrating device is the *chemical actinometer*, in which a certain amount of chemical 'turnover' is produced by a given number of incident photons. The chemical changes measured after a given time are proportional to the total radiation received. An example is the ferrioxalate actinometer, in which the radiation is absorbed by a few cm³ of potassium ferrioxalate in a glass or quartz cell. The light induces the release of ferrous ions into the solution, and the concentration of released ions is determined by adding a colorimetric reagent. The amount of colour produced depends on the concentration of ferrous ions and hence on the radiation received, and is measured optically. Typical exposures are of the order of an hour for the measurement of those light levels which can just be detected by thermopiles. The region of useful response extends from ultraviolet wavelengths up to about 500 nm in the visible spectrum.

7.5 Absorption of Light

We have already discussed the physics of the emission of light and seen that characteristic decreases in energy (transitions) of atoms or molecules can lead to the emission of a photon. All these transitions are, in principle, reversible, so that it is possible for the emission process to work backwards and for a photon of the correct energy to interact with an atom to produce an *increase* in the atomic energy, while the photon itself is absorbed. It follows that each type of energy transition in a material that could lead to emission at a characteristic wavelength can also absorb light at the same wavelength, since the energy change has the same value in either case. A striking example of this is the continuous spectrum of light from the sun, which is found to be crossed in places by spectral lines that are *dark*. These 'Fraunhofer lines' are due to the outer, cooler gases of the sun's surface absorbing more light than they emit, so that the sun's total emission is reduced at certain wavelengths. This absorption shows as the darker lines of the solar spectrum, which are found to correspond exactly in wavelength to the *emission* lines of common gases. Emission and absorption therefore both occur at wavelengths characteristic of a given substance, and analysis of the absorption spectrum will yield as much information about the material as its emission spectrum. In fact, the absorption spectrum is more useful in studying molecules of biological interest since it can be measured at room temperature, whereas the production of an emission spectrum usually leads to temperatures at which the molecules are degraded.

7.5.1 *Absorption in Optical Instruments*

Generally, in working with visible light, little notice is taken of the absorption of light by the liquids and optical components used in an instrument such as a microscope or spectrometer. This is because we use two good optical substances, glass and water, which have very low absorptions in the visible spectrum. Conventional glass transmits well between 450 nm and 1000 nm, while water transmits from the shortest ultraviolet wavelengths up to, again, about 1000 nm. However, when we wish to use infrared or ultraviolet light, absorption becomes a problem, and special optical materials, solvents, and techniques are necessary to work with these regions of the spectrum.

As a general principle, in the non–visible regions much more use is made of curved mirrors for focusing, rather than lenses in which the absorptive and refractive properties depend on wavelength. (Any mirrors that are used must, of course, be front–coated and not the type that has the reflective layer behind a sheet of glass.) Instead of using ordinary glass for the windows of instruments and for prisms in producing spectra, other materials must be found. For instance, fused silica glass can be used for both ultraviolet and infrared, since it transmits from 200 nm or less (u.v.) up to about 4 μm or 2500 cm^{-1} (near i.r.). Quartz optics are used predominantly in the ultraviolet, while in the far infrared one uses prisms made of a range of ionic crystals such as sodium chloride (rock salt) and caesium iodide. The latter transmits down to wavenumbers of 200 cm^{-1}. Since the use of prisms in the infrared is often inconvenient (some of the crystal materials are hygroscopic, have limited ranges of transmission or have undesirable refractive properties), it is common to find the prism replaced by a diffraction grating (Chapter 9) as a means of producing a spectrum.

To examine substances in solution, suitable solvents have to be employed which transmit at the wavelength of interest. In the ultraviolet, liquids such as water and various alcohols cut off below about 200 nm, while heptane transmits down to 170 nm. In the infrared region, carbon disulphide (CS_2) is a commonly used solvent.

In addition to the solid and liquid components of the system, an instrument usually contains air, which also has absorptive properties. Water vapour, for instance, has a range of absorption bands in the infrared spectrum from about 900 nm upwards, so that infrared devices should be used with dry air. At the other extreme, the oxygen of the air absorbs ultraviolet light below 200 nm, so an optical system is commonly filled with nitrogen in order to reach shorter wavelengths. However, at 170 nm nitrogen also cuts off, so that to reach still smaller values the apparatus must be evacuated entirely—the spectrum below 170 nm is for this reason called the 'vacuum ultraviolet' region.

7.5.2 The Effect of Absorption on Intensity

Suppose a beam of light of initial intensity I_0 (in units proportional to energy per unit area per second) passes through an optical component of thickness d and has afterwards an intensity I. We can describe the total amount of absorption that takes place by quoting the ratio of I/I_0, a quantity known as the *transmission* (τ) of the component. More usefully, we can calculate the *optical density* (O.D.) of the component, defined by

$$O.D. = \log_{10}\left(\frac{I_0}{I}\right) = \log_{10}\left(\frac{1}{\tau}\right). \tag{7.3}$$

(To give an example, if the component reduces the intensity by a factor of 100, then $\tau = 0.01$ or 1 per cent. and the O.D. $= \log_{10} 100 = 2$.) Obviously the absorption produced by a component of given material depends on its thickness, d, and the dependence is found to obey Lambert's law, which is

$$I = I_0 \, e^{-Kd}, \tag{7.4}$$

where the absorption coefficient, K, depends on the material and the wavelengths of the light concerned. The transmitted intensity I thus decreases exponentially with increasing thickness. If, instead of a solid component, we have a solution of absorbing molecules in a transparent solvent (Figure 7.12), then Beer's law states that, *for low concentrations and light of one*

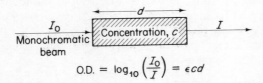

Figure 7.12.   Absorption by a solution of light of a given wavelength

wavelength, the factor K is proportional to the concentration, c, of the solution. We thus obtain for a solution in which the light has a path length d the equation known as the Beer–Lambert law:

$$I = I_0 \, e^{-kcd}, \tag{7.5}$$

where k depends on the solute and the wavelength. Writing the equation in another form, we have

$$\log_e\left(\frac{I_0}{I}\right) = kcd \tag{7.6}$$

or

$$\log_{10}\left(\frac{I_0}{I}\right) = \varepsilon cd, \tag{7.7}$$

in which $\varepsilon = 0.4343 \, k$. The left–hand side of equation (7.7) is the optical density of the solution, also known as the decadic absorbance or extinction, while ε is the extinction coefficient of the solute substance at a particular wavelength. (The units of ε depend on those of c and d; the usual combinations of c and d are mol m^{-3} and m or mol litre^{-1} and cm.) We can see from this equation that O.D. is proportional to the path length in the solution and to the concentration, so that measurements of O.D. at a given wavelength are a convenient way of estimating the concentration of a known substance.

7.5.3 *Absorption Spectrophotometry*

We can plot the emission spectrum of a light source by measuring the light intensity at many wavelengths. In a similar way we can plot an *absorption spectrum* by measuring values of K or ε for a material or solution. The absorption spectrum is characteristic of the substance and has the advantage that it can be obtained at room temperatures. It can either be used purely as a 'finger–print' to identify a particular compound or as a means of studying the molecular structure.

An instrument for measuring light intensities over a continuous range of wavelengths is called a spectrophotometer. The absorption spectrophotometer comprises essentially a light source, a device for isolating a small group of wavelengths from the source spectrum (a monochromator) and a detector (Figure 7.13). Light from the source passes through the mono-

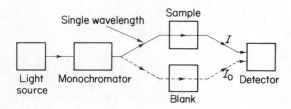

Figure 7.13.   Block diagram of an absorption spectrophotometer (single–beam type)

chromator, which contains a prism or diffraction grating, and from this a beam of light emerges which ideally has a single wavelength (monochromatic). The beam then passes through the sample to be investigated and finally falls on a detector. A method is necessary for comparing the final intensities of the beam, with (I) and without (I_0) a sample in the way. In the case of solutions, we want to find the effect of the solute, so the necessary comparison is between a transparent cell (a cuvette) containing solution and the same or an identical cuvette containing pure solvent (the 'blank').

In the simplest instruments, using a single light beam, the comparison is done manually by taking two successive detector readings, one for the sample and one for the blank. This is tedious, and in any case the light source may change in power between the two measurements, making them useless. In a double–beam spectrophotometer (Figure 7.14) two identical

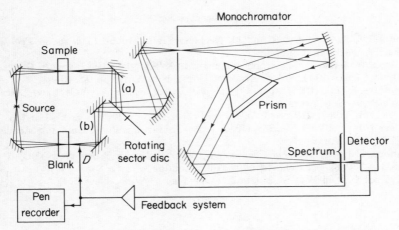

Figure 7.14. A simplified diagram of a recording double–beam spectrophotometer. The rotating sector disc alternately reflects beam (a) or allows the passage of beam (b) so that each in turn reaches the monochromator. Negative feedback is used to make (a) and (b) equal in intensity by controlling the diaphragm D

beams are taken directly from the source; one passes through the sample and the other through the blank. The two beams are then passed alternately through the monochromator to the detector by using a continuously oscillating mirror system, so that the detector receives an alternating signal unless the light from the blank cell (the 'reference' beam) is reduced mechanically, with a diaphragm, to the same power as that from the sample. The reduction is made automatically, using negative feedback (Chapter 4), until the fluctuating output of the detector vanishes. Then the mechanical reduction of the reference beam is just equal to the optical reduction produced by the sample in the other beam, and can be used to give a direct and automatic reading of optical density.

The advantages of using the reference beam system are that the detector receives chopped radiation, reducing the effects of drift and noise (Section 7.4.2), and that it is a *null method* (see the Wheatstone bridge and potentiometer, Chapter 4) and only depends on reducing the detector signal to zero; the way in which the detector responds to different intensities is of no consequence. Usually the optical density of the sample is displayed on a pen recorder, in which the paper speed is controlled by the movement of the prism or grating of the monochromator, so that a direct graph is recorded of absorption versus wavelength. The nature and some uses of these absorption spectra are described in the books mentioned below.

Further Reading

Bainbridge, R., G.C. Evans and O. Rackham (Eds.) (1966). *Light as an Ecological Factor*. Blackwell, London.
Driscoll, W.G. (1969). 'Spectrophotometers'. In R. Kingslake (Ed.), *Applied Optics and Optical Engineering*, Vol. 5. Academic Press, New York. pp. 85–104.
Kingslake, R. (Ed.) (1965). *Applied Optics and Optical Engineering*, Vols. 1 and 2. Academic Press, New York.
Rao, C.N.R. (1967). *Ultraviolet and Visible Spectroscopy*. Butterworths, London.
Seliger, H.H., and W.D. McElroy (1965). *Light: Physical and Biological Action*. Academic Press, New York.
Tanford, C. (1961). *Physical Chemistry of Macromolecules*. Wiley, New York.
Vasko, A. (1968). *Infra-red Radiation*. Iliffe, London.
Wyszecki, G., and W.S. Stiles (1967). *Color Science*. Wiley, New York.

PROBLEMS

7.1 A certain metal is said to have a 'work–function' of 2·0 electron–volts (i.e. the energy needed to eject an electron from the surface is equivalent to the energy needed to move the charge on one electron through a potential difference of 2 V). Calculate the value of the work–function in joules and then find the minimum frequency and maximum wavelength of light which would release electrons from the metal.

7.2 The radiation emitted by a certain molecular energy transition has a wavelength of 3 μm. Which kind of radiation is it, and what is the molecular process most likely to be involved? Calculate (a) the frequency of the radiation, (b) its wavenumber in cm^{-1} and (c) the corresponding transition energy in electron–volts (see problem 1 for definition).

7.3 In photosynthesis it is said that about nine photons are needed to convert one molecule of carbon dioxide to carbohydrate. If, for simplicity, we assume the light used has a wavelength of 663 nm, find the energy needed per carbon dioxide molecule. To what value does this correspond in kilo-calories per mole?

7.4 Suppose a photomultiplier has a noise–equivalent power of 10^{-16} W and is illuminated with green light of wavelength 500 nm. What is the minimum number of photons per second that would need to fall on the cathode for the light to be just detected? If each photon on average causes 10^6 electrons to be collected at the anode of the photomultiplier, what is the equivalent minimum anode current?

7.5 In a simple colorimeter the light transmitted by a 'blank' glass cell containing pure solvent gives an intensity reading of 96 units, while the intensity reading for an identical cell containing a coloured solution is 36·5 units. What is (a) the transmission of the solution and (b) its optical density? If the cell gives a reading of 59·2 units when filled with a diluted quantity of the same solution, what is the dilution ratio?

7.6 In an absorption spectrophotometer using cells of path–length 1·0 cm the optical density of a solution of protein is found to be 0·450 at a wavelength of 205 nm. The molar extinction coefficient of peptides at this wavelength is $250 \ m^2 \ mol^{-1}$. Find the concentration of the protein as moles of peptide per cubic metre.

CHAPTER 8

Optical Lenses

Lenses are used in a number of optical instruments, for example the microscope and the camera, which a biologist may employ in his laboratory. To make the best use of optical devices and to understand their limitations, it is essential to know how light behaves on passing through a lens. Lenses are made of materials which alter the direction of, or *refract*, a beam of light which enters the material obliquely to the surface. Glass is widely used in the manufacture of lenses intended for use with visible light, while quartz and rock salt may be employed for lenses to be used respectively for ultraviolet and infrared radiation. Certain types of clear plastic have recently been used successfully in the manufacture of special lenses and it is to be expected that this type of material will in the future enjoy a much wider application in this field.

8.1 Refraction

To understand how light behaves on passing through a lens it is necessary to discuss what happens to a beam of light on passing through a boundary which separates two different substances (e.g. glass and air). To do this we use *Huygens' construction*, in which each point on a wavefront of a light beam is regarded as a secondary light source which emits spherical waves. The position of the main wavefront at some later instant of time is found by drawing a surface which touches all the secondary wavefronts. For example, in Figure 8.1 the spherical wavefront generated by a point source of light is shown. Each point on this wavefront appears to act as a secondary source producing spherical wavelets, of which the profile, represented by the dotted line in Figure 8.1, is the new wavefront at a later instant determined by the velocity of light in the medium concerned.

Let us now apply Huygens' construction to a plane wavefront advancing obliquely towards a plane surface which separates two media (Figure 8.2). At some time the wavefront just touches the surface at O and makes an angle ϕ with it. We wish to find the position of the wavefront when the extremity of the original wavefront meets the surface at P, at a time t later. Since the light travels at different speeds in the two media, say v and v',

208

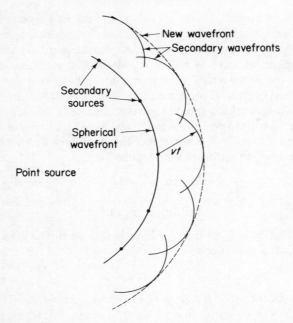

Figure 8.1. Huygens' construction used to find the position of a new wavefront at time t following the formation of a given wavefront. The velocity of light in the medium is v, so the radius of each secondary wavefront is vt

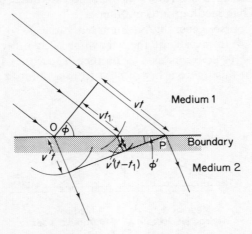

Figure 8.2. Huygens' construction used to find the position of a wavefront following refraction of light at a boundary separating two media. The velocity of light in medium 1 is v and in medium 2, v'

in a time t the distances travelled will be vt and $v't$ respectively. Thus if O is a secondary source on the wavefront it will produce a spherical wave-front of radius $v't$ in the new medium. A general point on the wavefront produces a spherical wave of radius vt_1 in the first medium and another, by regarding the intersection of sphere and surface as a further source, of radius $v'(t = t_1)$ in the second medium. The new wavefront is found by drawing the surface which touches all the spheres, a surface which turns out to be the plane passing through P and tangential to the sphere with centre O (Figure 8.2). This plane meets the boundary at an angle ϕ'.

We can see from the geometry of Figure 8.2 that

$$OP \sin \phi = vt \tag{8.1}$$

and

$$OP \sin \phi' = v't. \tag{8.2}$$

Now the velocity of light in a medium is related to the velocity of light *in vacuo* (c) by the expression

$$v = c/n, \tag{8.3}$$

where n is a constant called the *refractive index* of the medium. The refractive index of air is close to unity, that of water is about 1·3, while glasses have indices in the range from 1·5 to 1·6. From equations (8.1) to (8.3) we can obtain the expression known as Snell's law:

$$n \sin \phi = n' \sin \phi', \tag{8.4}$$

which can be used to determine the direction followed by a light beam on passing from one medium into another.

Lenses, of course, generally have curved surfaces, so we now make use of equation (8.4) to investigate the behaviour of light rays proceeding from one medium of refractive index n to another (n') across a spherical boundary (Figure 8.3). In this figure a ray of light from a point O strikes a spherical

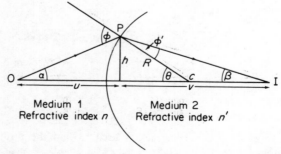

Figure 8.3.   Geometry of refraction at a spherical surface

surface (radius R, centre C) at P and is deviated or *refracted*. The angles ϕ and ϕ' between the normal to the surface at P (which is a radius of the sphere) and the incident and refracted beams are related by equation (8.4).

If we limit the discussion to small angles, so that the sine of each angle marked in Figure 8.3 may be set equal to the angle itself (in radians), then equation (8.4) becomes

$$n\phi = n'\phi'. \tag{8.5}$$

This restriction means that we are considering only rays close to the axis OI, the so–called *paraxial* rays.

In Figure 8.3 the following geometrical relations hold:

$$\phi = \alpha + \theta \tag{8.6}$$

and

$$\phi' = \theta - \beta \tag{8.7}$$

so that equation (8.5) may be rewritten as

$$n(\alpha + \theta) = n'(\theta - \beta). \tag{8.8}$$

Again, provided the angles are small,

$$\alpha = \frac{h}{u}, \ \beta = \frac{h}{v}, \ \theta = \frac{h}{R}, \tag{8.9}$$

so that equation (8.8) becomes

$$n\left(\frac{h}{u} + \frac{h}{R}\right) = n'\left(\frac{h}{R} - \frac{h}{v}\right). \tag{8.10}$$

This equation may be simplified and rearranged to yield

$$\frac{n}{u} + \frac{n'}{v} = \frac{n' - n}{R}. \tag{8.11}$$

Notice that this equation is independent of h, so all paraxial rays from O (at a distance u) will pass through the same point I (at a distance v). We say that light from the point O is *focused* at the point I by the surface. It is important, however, to note that this situation only applies for rays close to the axis of the system; in practical cases this is not always the case, as we shall discover later in this chapter.

If in Figure 8.3 a small object were placed at O, each point on the object would yield a corresponding point on the other side of the surface, thereby producing an *image*. There are few occasions when an image is formed in a medium different from that in which the object is situated, although important examples are the human eye and oil–immersion microscope objectives (Section 10.4). It is much more useful to have a device which will produce an image of an object in the same medium; such a device is a lens.

8.2 Image Formation by Lenses

A lens consists of some material (such as glass) enclosed by two curved surfaces, which are usually spherical but may have different shapes (e.g. cylindrical) in special cases. In particular, a lens surface may be convex, concave or planar so that several different types of lens are possible (Figure 8.4). We may consider that the image of an object, produced by the first

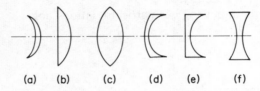

(a)   (b)   (c)   (d)   (e)   (f)

Figure 8.4.   Examples of lenses: (a) convex meniscus, (b) planoconvex, (c) biconvex, (d) concave meniscus, (e) planoconcave, (f) biconcave

surface of the lens, behaves as a virtual object for the second surface, which produces the final image. This situation is illustrated for a biconvex lens in Figure 8.5. Clearly, at each interface a ray of light must obey Snell's law,

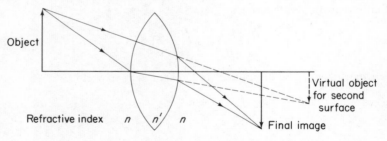

Figure 8.5.   Formation of a real image by a biconvex lens

so that by drawing rays from the object the position of the image can be established.

It is not usually necessary to adopt this procedure to find the position of the image, provided certain properties of the lens are known. Thus, in Figure 8.6 a ray of light (1) parallel to the lens axis passes through the *focal point*, F_2, of the lens. F_2 is the point at which a beam of light parallel to the axis would be focused. The projections of the incident ray and the ray through F_2 intersect in a *principal plane* P_2 of the lens, so that the incident beam appears not to be deviated until it strikes this plane. Similarly,

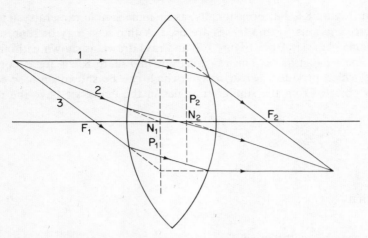

Figure 8.6. Illustrating the nodal points and principal planes of a thick lens

ray (3), which passes through the other focal point, F_1, of the lens, emerges parallel to the lens axis, and appears not to be deviated until it meets the other principal plane P_1. The points N_1 and N_2 at which the principal planes meet the axis are the *nodal* points of the lens, and have the property that a ray (2) directed towards N_1 (Figure 8.6) emerges from the lens as if from N_2, and is parallel to the incident ray. If, therefore, the positions of the principal planes and focal points are known, the image of an object can be constructed diagrammatically without drawing the lens surfaces; the lens need only be represented by its principal planes.

To obtain a simple relation between the image and object distances (u and v in Figure 8.7) and the focal length of the lens (f) we will consider the case of a thin lens in which the bounding surfaces are so close that the principal planes coincide. Most lenses used in practice are thick lenses,

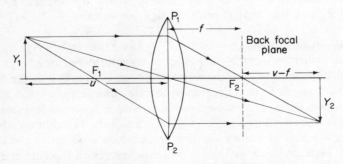

Figure 8.7. Construction for derivation of the thin lens equation

as in Figure 8.6, but occasionally, as in some simple camera systems, close approximations to thin lenses are used. A thin lens may be represented as a single plane (P_1P_2 in Figure 8.7) and rays drawn as shown without regard for the lens surfaces. The ray diagram in Figure 8.7 is for a converging lens which produces a real image (which can be shown on a screen) of a real object. From the similar triangles in the figure we have the relations

$$\frac{Y_2}{v-f} = \frac{Y_1}{f} \tag{8.12}$$

and

$$\frac{Y_2}{v} = \frac{Y_1}{u} \tag{8.13}$$

so that

$$fv = u(v-f). \tag{8.14}$$

This may be rewritten as

$$\frac{1}{u}+\frac{1}{v} = \frac{1}{f}, \tag{8.15}$$

which is the *lens equation* for a thin lens. The points at distances u and v from the lens are known as conjugate points. A somewhat more complex equation may be derived for a thick lens. Note that the magnification of a thin lens Y_2/Y_1 is equal to v/u from equation (8.13).

If the analysis given above is carried out for a thin biconcave lens, equation (8.15) is of the same form except that the signs of $1/v$ and $1/f$ are negative. It is clearly inconvenient to have two different formulae representing essentially the same problem. This situation is avoided by adopting one of several sign conventions. In one of these conventions, for example, equation (8.15) is used in all cases and distances to *real* objects and *real* images (i.e. images which can be formed on a screen) are regarded as positive while distances to *virtual* objects and images are regarded as negative. The focal lengths of convex and concave lenses are respectively positive and negative in this 'real–is–positive' sign convention.

8.3 Aberrations of Lenses

In the earlier sections of this chapter we considered only rays close to the axis of the lens, so that the angles between light rays and the axis were always very small. In practice, lenses can be quite large, so that rays some distance from the axis must be considered and in this case it is possible for some rays to make large angles with the lens axis. In addition, the transparent lens material has a refractive index which varies with the colour of the light incident upon it. Thus, if white light falls on the lens, the various coloured rays will be refracted by different amounts. These practical problems mean that a simple lens cannot produce a perfect image of an extended object, and the failings of the lens in this respect are called *aberrations*. We may conveniently consider the aberrations in two sections, the first dealing with defects which occur in monochromatic light and the second with imperfections in image formation associated with white light or light containing more than one colour. There are seven aberrations in all, known collectively as the *Seidel* aberrations.

8.3.1 *Monochromatic Aberrations*

Spherical aberration occurs for object points on the axis of the lens and occurs because the lens surfaces are spherical. The position of the image point depends on the distance from the axis at which a ray passes through the lens. As shown in Figure 8.8, light which passes through an outer ring

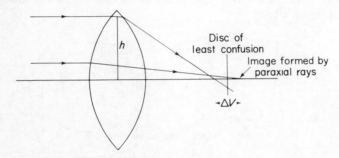

Figure 8.8. Spherical aberration for a convex lens. Light passing through an outer ring of the lens is brought to a focus closer to the lens than light passing through an inner ring

or *zone* of the lens is brought to a focus closer to the lens than a ray traversing an inner zone of the lens. The distance between the images formed by rays passing through the centre of the lens and through a particular zone is known as the longitudinal spherical aberration, Δv (Figure 8.8). To a close approximation Δv is proportional to the square of the radius h of the zone through which the rays pass to form an image. The ratio $\Delta v/v$, where v is the image distance for rays passing through the central region of the lens, is usually a few per cent.

It should be clear from Figure 8.8 that a screen placed in the region of the image will show a circular patch of light instead of the ideal point. The radius of the circle will change as the screen is moved, but at some point it will have a minimum value. This circle is called the *disc of least confusion*.

Spherical aberration cannot be eliminated for a single lens with spherical surfaces, but it can be minimized by choosing the shape of the lens carefully. It is found that the best shape from this point of view is intermediate between equiconvex and planoconvex, and is such that each surface refracts the rays more or less equally.

Two lenses with focal lengths of opposite sign can be combined to eliminate spherical aberration for one particular zone, since the spherical aberrations of the separate lenses compensate each other. It is, however, impossible to remove spherical aberration for all zones simultaneously unless the surfaces of the lens have non–spherical surfaces.

Oblique spherical aberration occurs for object points off the axis of the lens and occurs for the same reasons as axial spherical aberration.

Distortion is an aberration which occurs because the magnification produced by a simple lens depends on the distance of a point object from the axis ('object height'). When the magnification increases as the image height increases, *pincushion* distortion is seen in the image, while *barrel* distortion arises when the magnification decreases as the image height increases (Figure 8.9).

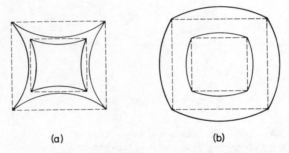

(a)   (b)

Figure 8.9. (a) Pincushion distortion; (b) barrel distortion. In each case the solid lines represent the actual image while the interrupted lines represent the 'correct' image

Coma is an aberration which exists for off–axis object points and occurs because each zone of the lens produces a circular image in a slightly different place from that of a neighbouring zone. The size of each circle varies with the radius of the zone, being larger for the outer regions of the lens. The image of a point object is simply the superposition of the separate images produced by each zone, and is a comet–like shape as shown in Figure 8.10. In practice, the image is more complicated because the lens usually suffers from spherical aberration as well as coma.

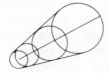

Figure 8.10. Showing how the comet–like image characteristic of coma is formed from a series of overlapping circles of light

Another off–axis aberration is that called *astigmatism*. Rays passing, for example, through tangential planes in the lens (the vertical planes in Figure 8.11) converge more strongly than rays passing through sagittal planes (the horizontal ones in the figure), with the result that two separate line

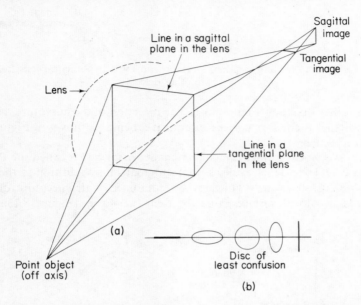

Figure 8.11. (a) Showing how the line images characteristic of astig–matism are formed. (b) Appearance of the image on moving a screen between the two line foci

images (one vertical, one horizontal) of a point object are produced. The appearance of the image as a screen is moved from one line focus to the other is shown in Figure 8.11b. The best representation of the object is the circular image which is known as the disc of least confusion.

It is, perhaps, interesting to note here that the astigmatism of a simple lens and the eye defect called astigmatism have different origins. Astigmatism of the eye arises from a distortion of the spherical surface of the cornea and affects images on, as well as off, the axis of the system. Astigmatism in a simple lens is an off–axis aberration which is a natural consequence of the spherical shape of the lens surface.

A further monochromatic aberration of a simple lens is *curvature of field*, a fault best illustrated by first considering the hypothetical case of a simple lens free from all the defects described above. Using the methods of ray–tracing described earlier, it is found that each point in a flat extended object is represented by a point in the image, but that the image itself is

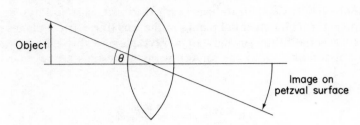

Figure 8.12.   Showing curvature of field for a hypothetical lens free
from other defects

curved rather than flat (Figure 8.12). The direction of the curvature depends
upon whether the lens is convex or concave, being as shown in Figure 8.12
for a convex lens.

The curved focal surface on which images are formed is called the Petzval
surface. A real simple lens suffers from astigmatism (in addition to the other
defects discussed earlier) which, when combined with curvature of field,
produces two focal surfaces, one for each line focus (Figure 8.13a). The

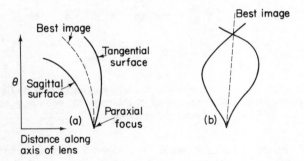

Figure 8.13.   Curvature of field for real lenses: (a) a simple biconvex
lens and (b) a Cooke triplet (see Figure 8.17c), showing how this defect
can be minimized. θ is the angle at which a ray passing through the
centre of the lens crosses the axis (see Figure 8.12)

separation of the surfaces for a simple lens increases as the angle (θ in
Figure 8.12) between the extreme light ray and the axis becomes larger.
Curvature of field can be reduced for a single lens by restricting the angle
of the cone of rays accepted by the lens. This is easily done by placing an
opaque screen (called a *stop*) with a hole of the appropriate size in front of
the lens. By using a combination of several lenses, curvature of field can
be reduced even when the angle between the extreme ray and the axis is
large. The lenses are then chosen so that the two focal surfaces cross at some
point and the best image lies in a plane (Figure 8.13b).

8.3.2 *Chromatic Aberration*

Because the refractive index of the lens material varies with wavelength, light of one colour is refracted more than that of another on passing through the lens so that an object illuminated by white light has many images (Figure 8.14). (This variation in refractive index is called *dispersion*, and

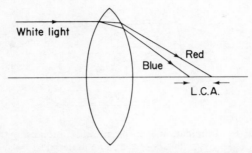

Figure 8.14.   Chromatic aberration. Blue light is brought to a focus closer to the lens than is red light. L.C.A. is the longitudinal chromatic aberration

is the characteristic of refracting materials which is responsible for the production of a spectrum by a prism.) The distance along the lens axis between images formed in red and in blue light is called the *longitudinal chromatic aberration*.

The focal points for light of two different wavelengths can be made to coincide by using two lenses of different material (and hence different dispersive powers) cemented together, or by using separated lenses of the same material. In the first case the two lenses must have curvatures of opposite sign so that their dispersions compensate (Figure 8.15), but overall

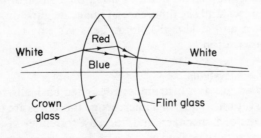

Figure 8.15.   Elimination of chromatic aberration using a doublet of lenses made from differing materials

deviation of a ray is still possible because different materials (e.g. crown and flint glass) are used for the two lenses. This type of compound lens is called an *achromatic doublet*, and is often used as the objective lens of a simple telescope.

When two separated lenses are used to compensate for chromatic aberration, the separation of the lenses must be equal to the average of their

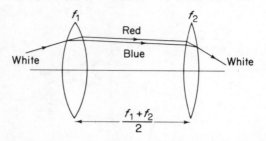

Figure 8.16. Elimination of chromatic aberration using two separated lenses of the same material

focal lengths (Figure 8.16). This type of combination is commonly found in the eyepieces of optical instruments.

The above combinations of lenses allow correction of chromatic aberration for two wavelengths only (usually those corresponding to red and blue light). Close examination of the image produced by such a lens shows that it is still coloured. Using lenses made of the mineral fluorite in combination with glass lenses, the chromatic aberration (and also the spherical aberration) can be made much smaller than is possible with a system composed wholly of glass lenses. Such a lens system is termed *apochromatic*.

In addition to longitudinal chromatic aberration, most lenses suffer from the defect known as *lateral colour*, which is an off–axis aberration. The magnification of a simple lens depends upon the colour of the light used to illuminate the object, and the difference in image heights for light of different colours is known as the lateral colour. Lens systems can be designed to minimize this aberration.

8.4 Design of Lens Systems

Lens systems are used in instruments designed basically to assist the eye in making observations. Thus, in a microscope the lenses are used to produce a magnified image of a specimen and, thereby, to render visible details that could not be seen with the naked eye. Again, a camera may be used to record events too rapid for the eye to follow or to photograph a specimen with intricate structural details which may then be studied at leisure. The lenses of these instruments must produce an image that bears as close a resemblance as possible to the object, although, in general, the use to which the lens system will be put requires only that it should reproduce the object faithfully under a limited set of physical circumstances. In a microscope objective, for example, good image quality is required only for one object distance and its corresponding image distance. Outside these conditions the image quality may be poor, since the lens is not intended for use there.

There are only a limited number of parameters which a lens designer can vary to modify the behaviour of a lens system and hence to bring it close to the required performance.

These parameters are called *degrees of freedom* and include such characteristics as the number of lenses in the system as well as the radii of curvature of the lens surfaces. The material of a lens can also be altered so that its refractive index and dispersive power change. In addition, the thickness of each lens can be modified, although this is not a useful method for eliminating the main aberrations, and stops can be placed at suitable points in the system. There must be at least eight degrees of freedom in a lens system if the seven Seidel aberrations are to be corrected. The word 'corrected' is used here in the sense of reducing to a predetermined maximum the value for a particular aberration; the maximum allowed for each aberration is governed by the use to which the lens will be put. As an example, three separated lens elements will provide the required number of degrees of freedom to correct for all the aberrations. The main degrees of freedom are the six lens surfaces and the two separations between the lenses. The material and thickness of each lens can also be modified to improve the performance of the system, but, in principle, the Seidel aberrations can be corrected by using three lenses all of the same glass.

Once the requirements of a lens system are known to the lens designer, the procedure to be followed is basically simple, although the details are very complicated. As a first step a trial system of lenses is set up on the basis of the designer's experience and ingenuity. This system is then evaluated and compared with the original requirements. If these are met no modifications are necessary; if not, design changes are made and the system re-evaluated. This procedure continues until a satisfactory system is attained. In principle, the lens designer evaluates his design by tracing the paths of a variety of rays through the system to investigate its behaviour. Clearly this procedure is very time–consuming and, since the system can be represented mathematically, great use of computers is made in the evaluation of lens systems.

8.5 Camera Lenses

It is instructive to discuss briefly the historical development of camera lenses as an example of how advances in knowledge and technique have produced improved designs. A normal camera lens (i.e. one intended for general photography) should produce an acceptable image of objects both on and off the lens axis, using light rays making angles of up to about 30°

with the axis. The images should be satisfactory for objects at a wide range of distances from the lens.

In the early part of the nineteenth century Wollaston produced the first useful lens for general photography. It was a convex meniscus lens combined with a small aperture stop a small distance in front of the lens (Figure 8.17a).

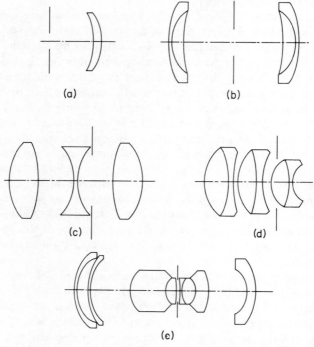

Figure 8.17. Showing how camera lenses have increased in complexity with development: (a) Wollaston meniscus, (b) rapid–rectilinear, (c) Cooke triplet, (d) Taylor–Hobson panchrotal, (e) Zeiss Biogon. (Not to scale)

This combination was found to be reasonably free from coma, astigmatism and spherical aberration provided the aperture in the stop was sufficiently small, although the system will obviously suffer from chromatic aberration. With this lens satisfactory images are formed from rays making angles of up to 20° with the axis. A development of the Wollaston meniscus lens was the rapid–rectilinear lens designed by Dallmeyer in the mid–nineteenth century. This lens, illustrated in Figure 8.17b, is symmetric about the central stop, a construction which renders it completely free from chromatic aberration, distortion and coma when used at unit magnification. Even when

not used under this condition the rapid–rectilinear lens is a great improvement on the Wollaston meniscus lens, although the aperture still has to
be small, so that the amount of light going through the lens is limited.
This situation improved following the production of a new glass having
a high refractive index but a low dispersive power, and a variety of symmetric
variations of the rapid–rectilinear lens were invented.

The next important development was the invention of the Cooke triplet
lens (Figure 8.17c) by Taylor in the latter part of the nineteenth century.
This lens was a departure from the symmetric systems used up to that
time and allowed rays at 30° to the axis to contribute to the image. The
Cooke triplet lens was the forerunner of many of the fine quality lenses in
use today, the Zeiss Tessar series being among the best known.

Modern photographers demand special lenses for some aspects of their
work. In bird photography, for example, it is desirable to use a telephoto
lens, which accepts only a narrow bundle of light rays and produces on the
photographic film a large image of a distant specimen. An example of such
a lens is shown in Figure 8.17d. Wide–angle lenses are important in certain
fields of photography (e.g. architectural and aerial survey work), since they
enable photographs to be obtained using rays that make very large angles
to the axis. In aerial photography a large area of land can be represented
on a single photograph if a wide–angle lens is used. The Zeiss Biogon,
a lens often used in aerial survey work, is illustrated in Figure 8.17e.

It should be evident from this brief survey how great have been the
advances in lens design in the last century or so. The advances have, of
course, been accompanied by improvements in the methods used to manufacture the lenses.

8.6 Microscope Objectives

An evolution of the kind described in the previous section has also occurred
in the design of microscope objectives. A microscope objective has different
design criteria from those of the camera lens. It will be used to form images
of points on or very close to the lens axis, so the system must be highly
corrected for axial aberrations. In addition it must accept rays making a
large angle with the axis to achieve good resolution (see Chapter 10). The
objective will also be used for a fixed object distance (and hence a fixed
image distance). The majority of routine microscope work is carried out
with standard achromatic objectives, an example of which is shown in
Figure 8.18a. For more demanding studies apochromatic objectives are
often used (e.g. Figure 8.18b). Both achromatic and apochromatic objectives
suffer from curvature of field which becomes pronounced at high magnifications. Objectives which produce a flat field are more complex than the
types so far discussed and are more expensive. At least one manufacturer

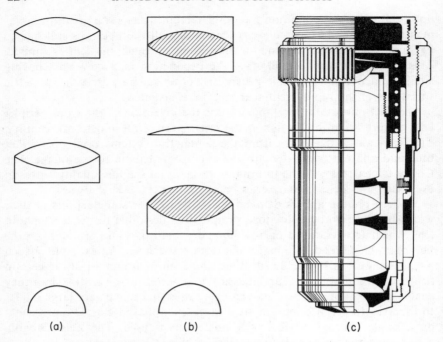

Figure 8.18.   Examples of microscope objectives: (a) a simple achromat, about ×40, (b) an apochromat, about ×50 (the shaded lenses are made of the mineral fluorite) and (c) a planapochromatic lens, ×100 (Figure (c) is reproduced by kind permission of Carl Zeiss, Oberkochen)

produces a range of flat–field apochromatic objectives, an example of which, showing the increased complexity, is shown in Figure 8.18c.

A further discussion of microscope objectives will be found in Chapter 10.

8.7 Aspheric Lenses

The component lenses of all the systems described in the previous sections are made of glass or fluorite and have spherical surfaces. It can be shown by the technique of ray–tracing that aberrations can be reduced by using lenses with aspherical (i.e. non–spherical) surfaces. To produce aspherical surfaces in glass lenses requires many hours of hand grinding and polishing, a prohibitively expensive procedure except in the most exceptional circum–stances. It is now possible, however, to obtain plastics of excellent optical quality from which to make lenses by a moulding procedure. This process, even for the production of non–spherical surfaces, is relatively inexpensive, particularly if large quantities of a given lens are required. At present these lenses are produced singly, and have not yet been exploited commercially in multicomponent systems, although aspheric plastic lenses are currently used in at least one inexpensive camera and also to magnify the map display in an aircraft navigational system. One drawback is that the plastic lenses are easily scratched, although this is being overcome by using surface layers of a hard plastic. This should not, in any event, be a serious drawback in the replacement of glass by plastic lenses in high quality systems, since the lens coatings used to eliminate surface reflections (Chapter 9) must also be treated with care. The cheapness of plastic lenses and their relative ease of manufacture, together with their good optical qualities, indicate that their use will increase markedly in precision optical equipment in future years.

Further Reading

Bracey, R.J. (1960). *The Technique of Optical Instrument Design.* English Universities Press, London.

Kingslake, R. (Ed.) (1965). *Applied Optics and Optical Engineering,* Vol. 3. Academic Press, New York.

Martin, L.C. (1954). *Technical Optics.* Sir Isaac Pitman, London.

Morgan, J. (1969). *Introduction to University Physics,* Vol. 2, 2nd ed. Allyn and Bacon, Boston.

Pitchford, A. (1959). *Studies in Geometrical Optics.* Macdonald, London.

PROBLEMS

8.1 A small fish 25 cm below the water surface is observed by a man whose eyes are vertically above the fish and 1·7 m from the surface. (a) How far below the surface does the fish appear to be to the observer? (b) How far above the surface do the observer's eyes appear to be to the fish? (Refractive index of water = 1·33.)

8.2 A camera has a thin lens which can be moved relative to the film to focus the image. The lens is 10 cm from the film when images of distant objects are sharply focused on the film. If the lens is moved 5 mm from this position, how near can the camera be placed to an object if the image is to be sharply in focus? What is the magnification in this position?

8.3 The simple eye of a certain spider has a spherical cornea of radius 0·1 mm and refractive index 1·3. The light–sensitive receptors are 0·44 mm from the cornea, while the material between the cornea and receptors has a refractive index of 1·3. How far is an object from the cornea if its image is in focus on the receptors?

8.4 A microscope eyepiece consists of two lenses of equal focal length separated by such a distance (0·05 m) as to compensate for chromatic aberration. The final image is virtual and is formed at a distance of 0·25 m from the lens nearest the observer. Determine (a) the focal length of each lens and (b) the overall magnification of the eyepiece.

CHAPTER 9

The Wave Properties of Light

When we say that light is electromagnetic radiation this means that it consists of oscillations in strength of both electric and magnetic fields. These oscillations are propagated through space in essentially the same way that oscillations in water level travel over the sea or oscillations of displacement travel up and down a guitar string. They each behave as waves and obey the laws of wave motion. The only essential difference is that waves in water or on strings imply actual movement, while light waves represent changing field strengths. Many of the most interesting properties of light are directly related to its wave–like character (the interference that produces colours in thin films, the diffraction that causes haloes to be seen round sources of light and the phenomenon of polarization), and the behaviour of light cannot really be understood without some knowledge of wave theory. To the scientist, the wave nature of light can be a nuisance, when it smudges the detail he wants to see with his optical instruments; but it can in other ways be an advantage, when for instance he uses the phase microscope or interference microscope, or investigates the structure of molecules with polarized light.

9.1 Behaviour of Transverse Waves

The examples of oscillation given above—light, water waves and waves on strings—are *transverse* waves, for the oscillations take place in a direction which is perpendicular to their direction of travel. Figure 9.1 shows the

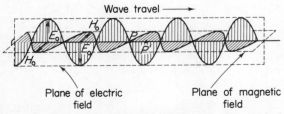

Plane of electric field

Plane of magnetic field

Figure 9.1.   Electric and magnetic fields of a plane–polarized ray of light

magnitude and direction of the electric field (**E**) and magnetic field (**H**) at every point along a single 'ray' of light, at a given instant of time, and we can see that both **E** and **H** are at right angles to the direction of the ray and, also, to one another. As the light waves pass any point in the path of the ray, the values of **E** and **H** at that point (e.g. P in Figure 9.1) grow from zero to a maximum, decrease to zero, grow again in the reverse direction, and so on. Corresponding events take place at another point such as P', but all at a later time than at P; the time interval between similar events occurring at P and P' obviously depends on the velocity of the waves. What is shown in Figure 9.1 is, in fact, *plane–polarized* light, for the electric field **E** remains always in a single plane (the vertical one in the figure). Ordinary, unpolarized light only differs from this in that it consists of a mixture of polarized light rays with the planes of their electric fields at all possible angles around the direction of propagation.

The energy carried by the waves along their direction of travel is proportional to the product of E and H at any point, but since H is always E multiplied by some constant value we can say that the energy is proportional to E^2. The average value of the energy in the waves will evidently be dependent on their size, a convenient measure of which is the peak value (E_0) attained by the field E. Thus we find that in a beam of light the average energy carried per second across a unit of cross–sectional area (the *intensity* of the light) is proportional to $E_0{}^2$ (the square of the maximum *amplitude*).

9.1.1 *Plane Waves*

In treating the theory of waves it is usual to simplify matters by considering

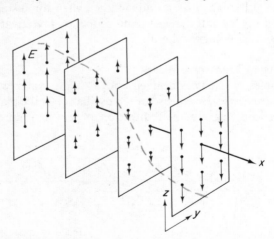

Figure 9.2. Electric fields in a plane–parallel wave travelling in the x–direction. All points in a given y, z–plane have the same field strength. (Magnetic fields are omitted for clarity.)

a beam of parallel light, travelling in a straight line along, for instance, the x–direction of a right–angled coordinate system. If all the electric (and magnetic) fields in the beam are 'in step' so that the values of E and H are constant over the whole of the y, z–plane that passes through a given value of x, then the beam is said to consist of *plane waves* (see Figure 9.2). The planes that pass through all points with the same E and H are called wave–fronts. At any point in the beam that is a distance x from the origin, the amplitude of the electric or magnetic field depends only on x and the time t at which the field is measured. The amplitude in the wave undergoes simple harmonic oscillations, that is it varies like a graph of $\sin \theta$, whether we plot amplitude against time for a fixed point x in space or consider the whole train of waves at a fixed instant t in time and plot amplitude against distance. (This can be appreciated by the analogy of wading in the sea— where one is standing the sea–level varies sinusoidally with time, but looking out to sea one observes that the level varies sinusoidally with distance.)

A graph of $\sin \theta$ is given in Figure 9.3, and it is apparent that by using suitable multiplying factors for the values on the axes this graph could be

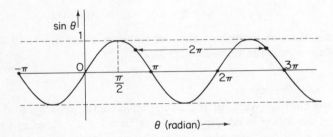

Figure 9.3.   Graph of $\sin \theta$ versus θ

stretched in either direction to represent the amplitude of a plane light wave plotted against either distance or time. Note that similar points on the $\sin \theta$ graph are separated by a value of θ equal to 2π. The behaviour of a plane wave that is travelling along the positive direction of the x–axis can be neatly described by a single equation as follows:

$$A = A_0 \sin\left[\frac{2\pi}{\lambda}(x-ct)\right], \tag{9.1}$$

in which A is the amplitude (value of E or H) at a given position x and time t. The quantity in the square brackets is effectively an angle of which the sine is taken, and since the sine of an angle does not rise above unity (Figure 9.3) the right–hand side of equation (9.1) has a maximum value of A_0. Thus, A_0 is the maximum amplitude of the wave. If we consider the amplitude A at a fixed position x, then x is constant in the square bracket but time t goes on increasing and A varies sinusoidally with t. Alternatively, if we consider the whole wave train at a fixed time t, the square bracket increases with x and A varies sinusoidally with x.

We now need to deduce the significance of the constants λ and c in the equation. Suppose, at a given time, we consider the amplitudes of the wave at a point x and a point $(x + \lambda)$. At the second point, inserting $(x + \lambda)$ instead of x, the square bracket is increased by $(2\pi/\lambda) \times \lambda$ relative to its value for the point x; in other words, the effective angle is changed by 2π. But the sines of two angles differing by 2π are the same (Figure 9.3), so the amplitudes of the wave at points x and $(x + \lambda)$ are equal. The constant, λ, is thus the distance between points of equal amplitude, which we call the *wavelength* of the wave.

Let us now consider a certain amplitude of the wave which occurs at given values of x and t. As time increases, how fast does that point on the wave move forward? To maintain A at a constant level the value of $(x - ct)$ in equation (9.1) must remain constant. If t increases by Δt, x must increase by $c\Delta t$ to achieve this, since in that case

$$(x + c\Delta t) - c(t + \Delta t) = x - ct = \text{constant.}$$

A constant level of amplitude thus advances a distance $c\Delta t$ in a time Δt, or, in other words, the *wave velocity* is c.

9.1.2 *Phase Difference*

Suppose another wave of the same wavelength, velocity and direction as the one described in equation (9.1) started off at a different time or from a different position from the original one. The equation describing this second wave would then differ from the first by having a constant time or constant distance added to the corresponding values in the square bracket. In either case the net effect would be to add some constant value to the value of the bracket, so that the equation would appear as

$$A' = A_0' \sin\left[\frac{2\pi}{\lambda}(x - ct) + \varepsilon\right], \tag{9.2}$$

in which A' is the amplitude of the second wave. The square bracket, which is effectively an angle, is usually called the *phase angle*, or just *phase*, of the wave and ε in this case is the *phase difference* between the waves described in equations (9.1) and (9.2). To give an example, if the wave A' were half a wavelength ($\lambda/2$) ahead of wave A, a constant value of $\lambda/2$ would need to be added to the values of x in equation (9.1) to make it describe the behaviour of A'. The phase of wave A' would thus be greater by $(2\pi/\lambda)(\lambda/2)$, that is π relative to the phase of A, so that the equation for A' could be written as equation (9.2) with $\varepsilon = \pi$. Half a wavelength difference between waves thus corresponds to a *phase* difference of π.

If the phase difference between two waves such as those described is zero or $2\pi n$, where n is some whole number, then for any chosen values of x and t the sine of the phase has the same value for each wave; both waves have their maxima at the same time and place, their zero amplitudes at the same time and place, and so on. The two waves are said to be *in phase* and their maximum amplitudes reinforce each other when the total electric or magnetic field is considered due to both waves (Figure 9.4a), provided that the two waves are polarized in the same plane.

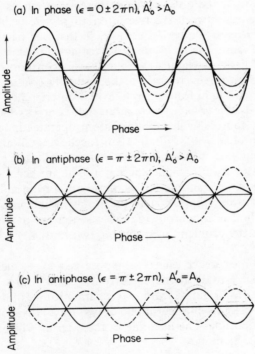

Figure 9.4. Addition of two similar plane waves: ——— A ------ A' ———
resultant. (a) In phase ($\varepsilon = 0 \pm 2\pi n$), $A'_0 > A_0$. (b) In antiphase ($\varepsilon = \pi \pm 2\pi n$), $A'_0 > A_0$. (c) In antiphase ($\varepsilon = \pi \pm 2\pi n$), $A'_0 = A_0$

If the phase difference between the waves is π (corresponding to half a wavelength in distance) or $\pi \pm 2\pi n$ (corresponding to an odd number of half–wavelengths), the sines of the corresponding phases at a given time and place are in exact opposition; if one sine has the value 1 (maximum) the other sine has the value -1 (minimum), and so on. The waves are said to be in *antiphase* and the amplitudes of the two waves tend to cancel each other out (Figure 9.4b). If both waves have equal maximum amplitudes the cancelling will be exact and the resultant of the two waves will be zero (Figure 9.4c).

9.2 Interference

The combination of two or more light waves, usually of the same wavelength, to give resultant effects of the kind shown in Figure 9.4 is called *interference*. When the waves are in phase so that their amplitudes reinforce each other, the interference is said to be *constructive*, and when the waves are in antiphase so that their amplitudes tend to cancel, the interference is *destructive*. For us to be able to see or detect the effects of interference, the phase difference between the two interfering wave trains must remain constant over a reasonable period of time, so that one particular resultant oscillation (e.g. Figure 9.4a or Figure 9.4c) remains in being for that period. If two sources of light give light waves that have a constant phase difference between them, the sources are said to be mutually *coherent*; such are the sources necessary for light from them to exhibit interference. Unfortunately, perhaps, it is impossible to have two separate light sources which are coherent, because in most sources each radiating atom operates independently and sends out a train of waves for about 10 ns (occupying about 3 m) followed by some random time interval before the next radiation. Thus, even the light waves emitted from a single source of light are generally incoherent, consisting as they do of random bursts of radiation from millions of atoms. Only in a laser source (Chapter 7) do the atoms radiate in phase, but there is as yet no method of synchronizing the waves from one laser with those from another, so the problem of producing two mutually coherent sources still remains.

Nevertheless, interference effects are observed experimentally. This is because it is possible to divide the light from a *single* source into two wave trains that appear to originate from two separate sources that are mutually coherent. The division is effected in one of two ways: division of wavefront or division of amplitude. The meaning of these terms will become clear as we discuss specific examples.

9.2.1 Division of Wavefront

A classic example of wavefront division is Young's experiment in 1801 which first demonstrated the interference phenomenon and the wave nature of light. In this experiment (Figure 9.5) the light source is behind a slit in a screen and acts as a line source of light. Cylindrical wavefronts spread out from the line and approach a second screen in which there are two parallel slits S_1 and S_2 a small distance d apart. If the two slits are equidistant from the source, parts of the same wavefront reach S_1 and S_2 simultaneously.

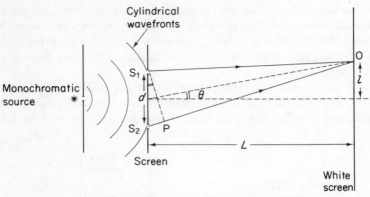

Figure 9.5. Young's slits experiment

Since all parts of a wavefront have by definition the same phase, the oscillations at S_1 and S_2 are also in phase. Now S_1 and S_2, by Huygens' principle (Chapter 8), act as new sources of light, and thus generate new cylindrical wavefronts which spread out towards the screen on the right of the figure. Thus, by selecting two parts of the same wavefront and allowing them to act as secondary sources, Young effectively created two sources of light which radiated coherently; this is the principle of wavefront division.

Let us now consider what happens at a point O illuminated by waves from the two sources. Obviously in the figure the light path S_1O is less than S_2O. The path difference, S_2P, causes waves from S_2 to reach O later than corresponding waves which originated simultaneously from S_1. If $S_2P = \lambda/2$, the two wave trains arrive at O with a phase difference of π, the oscillations at O are in antiphase and destructively interfere, and a dark horizontal line is seen across the screen at the position O. As we consider O to move further from the axis of the diagram (l increasing), S_2P increases to a value λ, the phase difference at O becomes 2π and constructive interference produces a bright line or fringe at O. As l increases further, more bright and dark fringes appear across the screen as the path difference passes through even and odd numbers of half–wavelengths respectively, and the screen is covered with an *interference pattern* of parallel fringes.

From the geometry of the diagram (Figure 9.5) we find that if $L \gg d$, $S_2P = d \sin \theta$ and $\sin \theta \simeq \tan \theta = l/L$. The path difference is thus given by $S_2P = dl/L$ and should be an odd number of half–wavelengths for destructive interference, that is

$$\frac{dl}{L} = (n + \tfrac{1}{2})\lambda \text{ for minimum intensity.} \tag{9.3}$$

Similarly,

$$\frac{dl}{L} = n\lambda \text{ for maximum intensity.} \tag{9.4}$$

Note, particularly, that at the centre of the pattern ($l = 0$) equation (9.4) is satisfied and there is a bright fringe, as one would expect, since O is then the same distance from S_1 and S_2 and the waves arrive in phase. Measurements of the fringe separations can be used to determine the wavelength of the interfering light.

There are several other methods for effecting a division of wavefront, but one which introduces a new phenomenon is Lloyd's mirror. In this arrangement (Figure 9.6) a slit source is used as in Young's experiment,

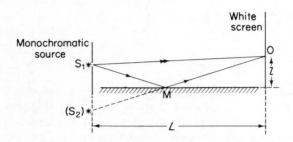

Figure 9.6.   Lloyd's mirror experiment

but the division of wavefront is produced by allowing part of the wavefront to be reflected by a mirror at almost grazing incidence. The reflected light appears to come from another virtual slit below the mirror and, again, we have two coherent light sources, one real and one apparent. The geometry of the system is exactly the same as in Young's experiment and the screen shows an interference pattern of horizontal fringes in the arrangement of Figure 9.6. However, there is one very important difference: where, from the geometrical considerations outlined for Young's apparatus, we would expect a dark fringe on the screen, in fact there is a bright one, and vice versa. For instance, if the mirror meets the screen, at their junction ($l = 0$) where a bright fringe would occur in Young's experiment, there is a dark fringe instead.

This reversal of the fringe pattern implies that the phase of the reflected light has been shifted so that, for instance, it now reinforces the direct beam reaching O instead of interfering destructively with it; a phase shift of π has occurred on reflection at the mirror. Such a phase shift occurs whenever light is reflected by a medium which has a higher refractive index than the medium in which the light is travelling, and effectively shifts the reflected wave train by half a wavelength relative to the position we would expect from purely geometrical calculations. (The sense of the shift, forwards or backwards, has no relevance because either has the same effect on any observed interference pattern.)

9.2.2 *Division of Amplitude*

An example of amplitude division is shown in Figure 9.7, where a cover slip inclined slightly on a microscope slide is illuminated from above by a parallel beam of light as shown. (The inclination can be produced by putting thin paper under one edge of the cover slip.) A half–silvered mirror enables

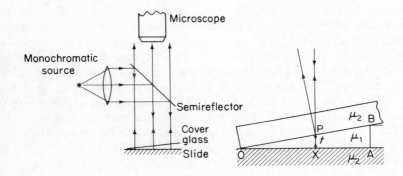

Figure 9.7.   Interference in a wedge–shaped film

light reflected from the system to be observed through a microscope. The space between the cover slip and the slide may be filled with air or some liquid of refractive index μ_1, but in either case the glasses enclose between them a film of wedge–shaped cross–section. Now the boundary OB between the lower surface of the cover slip and the wedge film is a partially reflecting surface so that when a wavefront from the source strikes it, the wave is both transmitted into the film and reflected upwards by the surface. The amplitudes of the transmitted and reflected versions of the wavefront are together, of course, equal to that of the original one. A single wave train arriving at OB is thus divided into two coherent wave trains of lesser amplitude, one reflected and one transmitted. The division of amplitude has thus created the necessary conditions for subsequent interference.

The transmitted light passes on through the thickness t of the film, is reflected at the surface of the microscope slide, passes again through the film and eventually enters the observing microscope. The light reflected at OB enters the microscope without passing through the film, although in other respects it follows a similar path to the transmitted light. The waves transmitted into the film thus lag behind those reflected at the top surface, and when the two wave trains ultimately interfere at the retina of the eyeball, the point X appears light or dark depending on the *optical path difference* between the two light paths.

As we have already seen in the case of Lloyd's mirror, the path difference that governs interference is not just the geometrically measured value, as we have to take into account phase changes on reflection (in this case at X, if $\mu_2 > \mu_1$ in Figure 9.7). We also have to take account of the speed with which the light travels over the extra path involved. As an analogy, imagine two cross–country runners side by side on the road and suppose A takes the wrong route into a ploughed field, discovers his mistake, and returns to the point where he left the road. His competitor, B, on the road is now ahead of A by a distance *further* than the distance A covered in the field, because the time lost by A in his detour is equivalent to one distance running slowly through the field, but a much greater distance running easily on the road. Similarly in the optical case, a path difference in a medium of high refractive index (low wave velocity) is equivalent to a greater path difference in vacuum, and it is customary in all interference calculations to convert all delays of one wave relative to another (including phase shifts) into the equivalent path difference in vacuum, the *optical path difference.*

Let us now calculate the optical path difference in the wedge film, which has a refractive index of μ_1. The velocity of light in the film is c/μ_1 and the time spent by the light in traversing the distance t twice is $2t/(c/\mu_1) = 2\mu_1 t/c$. During this time, light in a vacuum, with velocity c, would travel a distance $c.2\mu_1 t/c = 2\mu_1 t$. (We have here a general result, that the optical path in a medium is the geometrical path multiplied by the refractive index.) In addition, the light suffers a phase change of π on reflection at X if $\mu_2 > \mu_1$, which is equivalent to an optical path difference of $\lambda/2$. The total optical path difference Δ between the two wave trains which reach the eye is thus given by

$$\Delta = 2\mu_1 t + \lambda/2. \tag{9.5}$$

For the waves to interfere destructively so that a dark fringe appears at X, we have

$$\Delta = (n+\tfrac{1}{2})\lambda, \tag{9.6}$$

that is

$$2\mu_1 t = n\lambda. \tag{9.7}$$

The film is thus crossed by dark fringes (Figure 9.8) wherever the optical

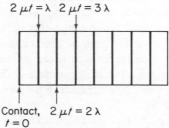

Figure 9.8.   Appearance of wedge–shaped film from above

path in the film becomes equal to a whole number of wavelengths. The fringes thus represent contour lines of constant optical path and indicate how the wedge increases in thickness from left to right.

One of the reasons why we have looked in detail at the wedge film is that it demonstrates the principle on which the interference microscope is based (Chapter 10). The contour fringes seen in the wedge are contours of *optical* path, $2\mu_1 t$, and will thus be affected by changes in refractive index in the film as well as changes in thickness. If a transparent cell body which has a refractive index different from that of the film is present under the cover slip, the fringes running across the cell will be displaced relative to those in the background, because the cell has modified the optical path (Figure 9.9).

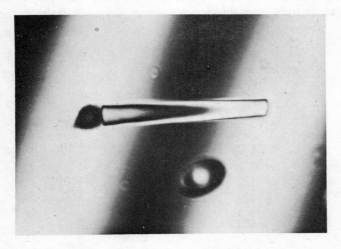

Figure 9.9.   Modification of the optical path by an object within a wedge–shaped interference system. A retinal rod from a frog retina is crossed by two bright fringes in an interference microscope. The bright fringe on the right–hand end of the rod corresponds to the left–hand bright fringe of the background, and has been displaced because of the extra optical path introduced by the rod. (Illustration kindly supplied by Dr. H.G. Davies, M.R.C. Biophysics Unit, King's College, London)

Interference can thus show the presence of transparent objects which would be almost invisible in a conventional microscope.

Similar fringes to those formed in a wedge–shaped film are produced in any film with partially reflecting surfaces, although their visibility depends on the nature of the illumination and the reflectivity of the surfaces. If a film is illuminated with white light, coloured fringes are seen, as colours of different wavelengths satisfy the conditions for constructive interference at different parts of the film or at different angles of illumination. The contour fringes of the type we have discussed are used in testing the regularity of lenses by examining, with a system of the kind shown in Figure 9.7, the air film between a lens and a piece of optically flat glass on which it is placed. In this case, if the lens has a lower surface which is perfectly spherical, the contour fringes (Newton's rings) are perfect circles.

An important application of the interference produced by amplitude division is in the construction of antireflection coatings of 'bloomed' lenses and of interference colour filters and dichroic mirrors. In the previous example of a wedge film we assumed that light was reflected only once from the back surface of the film, which is a fair assumption if the reflectivity of the film surfaces is low. However, if a thin film is constructed with highly reflecting surfaces, the light transmitted into the film 'ricochets' between them and *multiple–beam interference* occurs. Figure 9.10 shows the incident

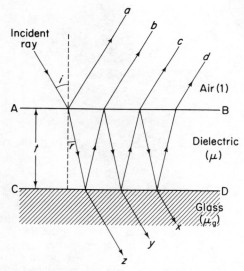

Figure 9.10.   Multiple–beam interference in an antireflection lens coating $(1 < \mu < \mu_g)$

light approaching such a film at a slight angle to the normal so that the multiple reflections are apparent in the diagram. Each time a light ray meets a surface, some of the light is reflected and some transmitted. (In Figure 9.10 is shown a single layer of dielectric material, such as magnesium fluoride, evaporated on the glass surface of a lens; in interference filters a thin semireflecting surface of silver or aluminium is often used at AB and CD.)

To discover what happens to the light reflected back into air, composed of rays *a*, *b*, *c*, etc., we consider the optical path difference between ray *a* and ray *b*. In the former the light has undergone one reflection at surface AB while in the latter the light has traversed the dielectric twice and has been reflected once at CD. Thus for ray *a*, the optical path difference Δ_a relative to the incident ray, because of the phase change on reflection, is given by

$$\Delta_a = \lambda/2 \qquad (9.8)$$

for a ray of wavelength λ. The corresponding optical path difference Δ_b is given by

$$\Delta_b = 2\mu\left(\frac{t}{\cos r}\right) + \frac{\lambda}{2}, \qquad (9.9)$$

since we have two paths of length $t/\cos r$ through the dielectric of refractive index μ, and a reflection at CD which gives a phase change of π if $\mu < \mu_g$. For rays *a* and *b* to interfere destructively (i.e. to minimize the intensity of reflected light) we require the optical path difference between the rays to be an odd number of half–wavelengths, that is

$$\Delta_b - \Delta_a = (n+\tfrac{1}{2})\lambda. \qquad (9.10)$$

For rays near normal incidence, $\cos r \simeq 1$ in equation (9.9) and the condition (9.10) becomes

$$2\mu t = (n+\tfrac{1}{2})\lambda \quad \text{for destructive interference.} \qquad (9.11)$$

It can easily be shown that for the same condition, the next pair of rays, *c* and *d*, also suffer mutual destructive interference, and so on. This means that, for a given wavelength, the thickness *t* of the film can be chosen so that equation (9.11) holds and there is a minimum of light intensity reflected from the dielectric film. In antireflection coatings *n* is usually chosen to be zero and thus $\mu t = \lambda/4$; λ is chosen to be in the middle of the visible spectrum (green) so that a range of wavelengths on either side approximately satisfies the condition for minimal reflection. The light that is reflected, at the two ends of the visible spectrum, appears with the complementary colour to green (i.e. purple)—hence the purple colour of bloomed camera lenses. It is left as an exercise to the student to show that in Figure 9.10, at a wavelength where *a* and *b*, *c* and *d*, etc., interfere destructively the rays *z* and *y*, *x* and *w*, etc., interfere constructively—minimum reflection implies maximum transmission.

9.3 Diffraction

If we isolate part of the wavefront from a source of light, for instance by allowing some of the light to pass through an aperture in a screen, the behaviour of the light on the further side of the screen can be predicted by using Huygens' principle. Each small part of the wavefront that passes through the aperture is considered as a secondary source of light, radiating in phase with all the other parts. The interference between the light radiated from all the different parts of the wavefront is called *diffraction*; this results, as we shall see, in a quite different distribution of light on the far side of the aperture from what we would expect if light were like pellets fired from a gun. There are two kinds of diffraction phenomenon, one called Fresnel diffraction which is concerned with curved wavefronts (spherical or cylindrical) and the other called Fraunhöfer diffraction which results from plane wavefronts and is consequently easier to interpret mathematically.

9.3.1 *Fraunhöfer Diffraction at a Narrow Slit*

Plane wavefronts arise in beams of parallel light and thus Fraunhöfer diffraction is observed when parallel light passes through a small aperture, as shown in Figure 9.11. Here, monochromatic light from an ideally point

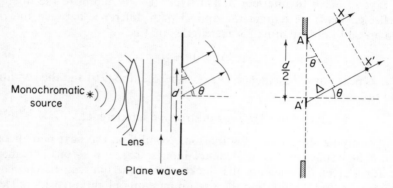

Figure 9.11.   Fraunhöfer diffraction at a single slit

source is focused with a lens to give plane wavefronts which approach a narrow slit of width d. The part of the wavefront that emerges from the slit is considered to radiate light according to Huygens' principle. As a mathematical convenience we divide the width of the wavefront into two halves and consider corresponding equal portions such as A and A' of each half as shown, which have a separation of $d/2$. The two portions act as Huygens' secondary sources of equal intensity and radiate cylindrical wavefronts (since they are portions of a narrow slit and act as line sources). Now if we simply consider rays which radiate at an angle θ to the forward path of light through the slit, the optical path difference in air between light which reaches a point such as X from A and that which reaches X' from A' is given by

$$\Delta = \frac{d}{2} \sin \theta. \tag{9.12}$$

If such a pair of rays is focused on a screen by a lens or is focused by the eye, destructive interference occurs first when Δ reaches a value of $\lambda/2$, that is

$$\frac{d}{2} \sin \theta = \frac{\lambda}{2} , \qquad (9.13)$$

or

$$d \sin \theta = \lambda \quad \text{for destructive interference.} \qquad (9.14)$$

Since the section of wavefront adjacent to A and the corresponding section adjacent to A' are also separated by $d/2$, similar reasoning applies to rays from them at an angle θ, and so on for each pair of sections of the wavefront passing through the slit. Thus, when equation (9.14) holds, any light intensity radiated from one section of the wavefront is cancelled by light radiated from another section $d/2$ away, and at the angle θ no light is received from the slit at all.

Now suppose we regard the slit of width d as a number (n) of slits imme–diately adjacent to each other, of width d/n, so that the width of all the hypothetical slits packed together is that of the original slit. For each slit of width d/n the direction in which destructive interference occurs is given by equation (9.14), in which we replace d by the width d/n of the slit concerned. Thus,

$$\frac{d}{n} \sin \theta = \lambda \qquad (9.15)$$

or

$$d \sin \theta = n\lambda \quad \text{for destructive interference.} \qquad (9.16)$$

Since all the hypothetical slits have the same width, d/n, equation (9.16) applies to all and no light is received from the collection of slits (i.e. the real

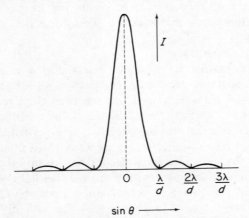

Figure 9.12.   Diffracted intensity from a slit, plotted against $\sin \theta$

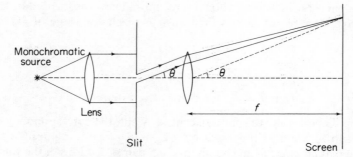

Figure 9.13.   Production of a Fraunhöfer diffraction pattern
on a screen

one) when this equation is fulfilled. This is a more general equation than equation (9.14) and shows that many angles are possible at which destructive interference occurs, corresponding to the various values of the integer, n. Obviously at intervening angles constructive interference occurs, so that if the intensity of light from the slit is plotted against angle the graph is as shown in Figure 9.12. The reality of this diffraction can be demonstrated by the method of Figure 9.13, in which a lens is used to focus all light at a given angle from the slit into a line on a screen in the back focal plane of the lens. Since increasing angles θ give alternately destructive and constructive interference, the screen is crossed by a *diffraction pattern* of dark and light fringes, instead of showing a sharp image of the slit. The intensity of the bright fringes decreases with angle, as shown by the maxima in Figure 9.12, but the theory of this need not concern us here. One point to note is that from equation (9.16) (illustrated by Figure 9.12) the larger the dimension of the slit, the smaller becomes the value of θ for a given dark fringe, so that a wide slit gives a diffraction pattern of narrow fringes and vice versa; this inverse relationship is characteristic of all diffraction patterns.

9.3.2 *The Diffraction Grating*

The diffraction grating is a useful optical device which provides an alternative to the prism for producing spectra from white light. In its simplest form it consists of a glass plate in which has been cut a regular pattern of parallel fine lines, of which there may be as many as 10^4 in a centimetre. The 'grating' thus created acts as a regular array of slits, each separated from its neighbour by a distance D (Figure 9.14) which is called the 'grating constant'. Each 'slit' acts as a separate line source of light, and when the diffraction grating is illuminated with plane parallel light of a single wavelength λ, the light radiated at a certain angle from the grating interferes constructively when focused. In Figure 9.14 the path difference between a ray from one source and one from its nearest neighbour (both at an angle θ

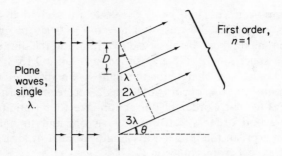

Figure 9.14.   First–order diffraction from a diffraction grating

to the normal) is $D \sin \theta$. Constructive interference will occur when the path difference is an integral number of wavelengths, so that rays from all the sources are in phase when focused, that is

$$D \sin \theta = n\lambda \quad \text{for maximum intensity.} \tag{9.17}$$

In Figure 9.14 the situation is shown when $n = 1$ (first–order diffraction), and rays from successive lines of the grating increase in path length by one wavelength at a time. Obviously a second maximum occurs in the diffraction pattern from the grating when $\sin \theta$ is twice as large and the path differences shown in the figure become 2λ, 4λ and 6λ, etc.; this is the second–order maximum, with $n = 2$.

The number of orders of diffraction that can be observed is limited because $\sin \theta$ in equation (9.17) can only increase as far as unity (for rays at grazing angle to the plane of the grating), and for a given grating constant D and wavelength λ only a few values of n and $\sin \theta$ may be possible. For example, a typical grating has 15 000 lines per inch which corresponds to $D = 1.69\ \mu\text{m}$. If green light with $\lambda = 0.546\ \mu\text{m}$ is used, from a mercury lamp, the first-

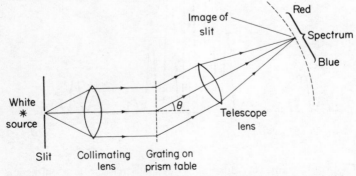

Figure 9.15.   Observation of a diffraction–grating spectrum with a prism spectrometer

order maximum occurs at $\sin \theta = \lambda/D = 0.323$, the second at $\sin \theta = 0.646$, and the third might just be observed at $\sin \theta = 0.969$ (a grazing angle). The diffraction maxima can be observed by mounting the diffraction grating vertically on a spectrometer table as shown in Figure 9.15. If a white light source is used, each wavelength of the light has a different value of θ at which maximum intensity occurs, and thus a spectrum is produced over a range of angles for each order of diffraction. We thus obtain a first–order spectrum, a second–order spectrum, and so on, and in contrast to the spectrum from a prism the longer wavelengths appear at larger values of θ, as one would expect from equation (9.17).

We have so far only discussed the *geometry* of the diffraction, that is how the angles at which maxima occur depend on the spacing between the lines of the diffraction grating. The *intensities* of the diffraction maxima depend on other factors such as the width of the transmitting portions of the grating and the total number of them present (i.e. the number of lines in the grating). Up to now we have neglected the fact that the transmitting element of the grating, which acts as a fine slit, has a finite width and is not infinitely narrow. However, the light from any slit of finite width behaves in the manner described in Section 9.3.1 if illuminated with parallel light, so that each element of the diffraction grating can be thought of as originally giving a Fraunhofer diffraction pattern as shown in Figure 9.12.

When many elements are combined to form a diffraction grating, the light from each being distributed in a Fraunhofer diffraction pattern, the maximum amplitude that can occur at a given angle in the pattern from the

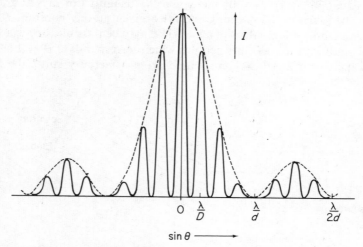

Figure 9.16.   Intensity distribution in the pattern from a diffraction grating where $D = 4d$. The dotted line is the shape of the pattern from a single element, with secondary maxima exaggerated for clarity

grating is obviously the sum of the amplitudes given at that angle by each element. Thus at an angle where the *single* element would give a *minimum* in its diffraction pattern, the diffraction pattern due to all the elements combined also shows a minimum. This is so even though equation (9.17) may predict that the amplitudes from all the elements add constructively— the sum of many zeros still remains zero. In a similar manner, where the single element would give a *maximum* in its diffraction pattern, the maxima of the pattern from the whole diffraction grating are greater than elsewhere. In summary, the result of these effects is that the intensities of the pattern from a diffraction grating are governed by the diffraction pattern due to one diffracting element, as shown in Figure 9.16. This shows the intensities when the spacing D between 'slits' in the diffraction grating is exactly four times the width d of a single slit. The maxima occur in the positions predicted by equation (9.17), but their intensities are controlled by the shape of Figure 9.12. This kind of phenomenon, where the intensity distribution of one diffraction pattern is controlled by the intensity curve of another pattern, is called convolution and is important in the interpretation of X–ray diffraction patterns.

One further point to note is that the sharpness of the diffraction peaks in Figure 9.16 (i.e. the sharpness of the fringes in the diffraction pattern) depends on the *total number* of diffracting elements present in the grating. A wide diffraction grating containing many lines will give a pattern with very fine, narrow fringes or maxima, and if the grating is used to produce a spectrum (Figure 9.15) the spectral lines will be sharp.

9.3.3 Fresnel Diffraction at a Circular Aperture

In the previous section we considered the diffraction of plane waves (Fraunhöfer diffraction) at a slit, but since it is common in optical instruments to have diverging beams of light and circular apertures we shall describe this case, although we shall not treat it mathematically. When light diverging from a point source meets a circular aperture, such as the stop in front of a lens, a circular area is selected by the aperture from the spherical wavefront surface that approaches it. In Section 9.3.1 we considered the plane wavefront penetrating a slit as a number of long, narrow sources of light. In the same way, we can think of the spherical wavefront in the present case as a number of annuli or *zones* (Figure 9.18a), each of which radiates light according to Huygens' principle. The interference between light radiated from the different zones produces a diffraction pattern of circular fringes around a central spot of maximum intensity, as shown in Figures 9.17 and 9.18. Since all lenses have a finite aperture this means that the image produced by a lens of a point source of light is always a circular diffraction pattern, never a sharp point. The angular spread of the pattern depends on the size of the aperture and the wavelength of the light, as in all diffraction patterns, and

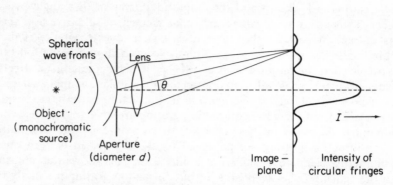

Figure 9.17.   Diffraction by a circular aperture (diffraction angles
exaggerated for clarity)

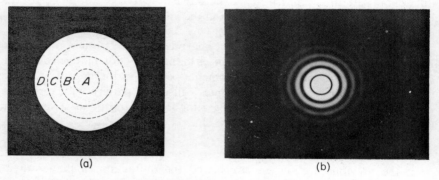

Figure 9.18.   A circular aperture can be divided, for calculation, into
zones (a) from which light interferes to produce an Airy type of diffraction pattern (b)

obviously causes a deterioration in the quality of any image produced by
the lens, since any point in the object is smeared out into a disc shape.
This bright disc, which corresponds to the central maximum of the diffraction
pattern, is called an Airy disc after its investigator. The diffraction pattern
produced when plane wavefronts are diffracted by a circular aperture is
called an Airy diffraction pattern and the angular positions of the maxima
and minima have been calculated. The first dark fringe, or minimum,
that surrounds the central spot of the pattern occurs at a radius that subtends
an angle $\theta_1 = 1 \cdot 22 \, \lambda/d$ at the diffracting aperture. Here λ is the wavelength
of the diffracted light and d the diameter of the circular aperture. Since
θ_1 gives a measure of the scale of the pattern we can see that the smaller
the aperture (d) becomes, the larger is the image disc corresponding to a

given point in the object, and the poorer the quality of the image. Although the quoted value of θ_1 is that for Fraunhöfer diffraction, values of a similar order occur in the Fresnel case.

9.3.4 *The Effect of Diffraction on Resolution in Optical Instruments*

In an optical system the image of a point object is generally not a perfect point because of defects in the lenses. Two bright points close together in the object may thus give two overlapping patches of light in the image, rather than two separate points. At separations less than some critical value the two object points cannot therefore be distinguished from each other; this critical value is called the *resolution* of the optical system and is the practical limit to the details which can be detected in the object. The resolution is to some extent dependent upon the contrast in the image as well as upon the optical system, but a working criterion due to Rayleigh is usually accepted when diffraction at the lens aperture is the main factor governing the image quality. Since the images of two close object points are then two overlapping Airy discs, Rayleigh suggested the central maxima of the two discs would just be distinguishable as separate (i.e. would just be resolved) if the maximum of one disc fell at the first minimum of the other (Figure 9.19).

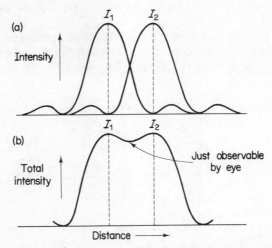

Figure 9.19.   Two Airy discs are just resolved if the maximum of one falls on the minimum of the other (Rayleigh criterion). (a) Intensity versus radius for two Airy patterns with centres at I_1 and I_2. (b) Total intensity along a line passing through I_1 and I_2

If the Rayleigh criterion is applied to the image formed by a single lens of two point sources of light, we obtain the situation shown in Figure 9.20.

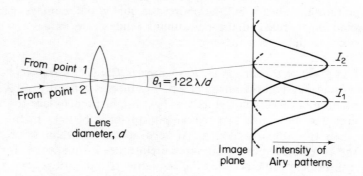

Figure 9.20. Angular resolution of a lens for two luminous object points

The point sources give rise to Airy discs in the image plane, and since the radius of the first minimum of the pattern subtends an angle $\theta_1 = 1\cdot22\ \lambda/d$ at the lens, this is the angle between the centres of the two patterns when the centre of one falls on the minimum of the other. The angular separation at the lens of two images that can just be resolved is thus $1\cdot22\ \lambda/d$ radian. If the point sources are two stars and the lens in this case is the objective lens of a telescope, it can be seen that the smaller the angle between the two stars, the greater must be the diameter (d) of the lens in order to distinguish them.

Although the Rayleigh criterion can be applied approximately to determine the resolution of a microscope, a more instructive approach is that of Abbe (the so–called Abbe theory). In this (Figure 9.21) we consider an object

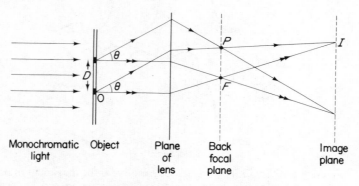

Figure 9.21. Abbe theory of image formation by a lens

under the microscope illuminated by monochromatic plane–parallel light to be a form of diffraction grating, where the various features in the object correspond to so many slits (or holes) in the grating. The problem, then, is to find the smallest distance, D_{min}, between slits in the grating that allows the slits to be resolved in the microscope image. The distance D_{min} is a measure of the detail which the microscope can reveal in a real object.

Now a perfect lens focuses all parallel rays to a point in its back focal plane and, in particular, all the rays diffracted at a given angle θ from the object (Figure 9.21) will pass through a given point P in this plane before proceeding to their destined points in the image. The back focal plane thus contains the *diffraction pattern* of the object; if a screen were placed in this plane it would reveal the pattern, in the same way that the pattern is produced in Figure 9.13. The point P in the back focal plane which corresponds to a diffraction angle θ will receive light if there is a spacing D between features of the object such that

$$D = \frac{\lambda}{\sin \theta} \, . \tag{9.18}$$

(This follows from the theory of the diffraction grating, Section 9.3.2). Another point in the diffraction pattern, corresponding to an angle θ^1, will receive light if, in the object, there is a spacing D^1 given by $D^1 = \lambda/\sin \theta^1$. Thus, each part of the diffraction pattern in the back focal plane of the lens is linked with a certain spacing between the features of the object. Information about the presence of large spacings is carried near the centre of the pattern, where $\sin \theta$ is small, while information about small spacings is present in the outer regions of the pattern, where $\sin \theta$ is large. This is shown in Figure 9.22, where D is plotted against $\sin \theta$, using equation (9.18).

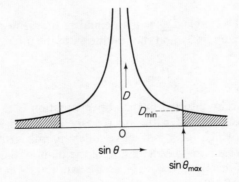

Figure 9.22. Graph of $D = \lambda/\sin \theta$ to show how, in a diffraction pattern, information about large values of D is carried at small values of θ

Since the light rays that form the image (Figure 9.21) are the same ones that formed the diffraction pattern, it follows that all the information neces-sary to reproduce the image must be carried, in another form, in the diffraction pattern. What happens when we prevent all the information of the diffraction pattern from reaching the image is shown by the shaded areas in Figure 9.22. If an aperture is placed in the optical system so that rays with initial angles greater than a certain value, θ_{max}, are prevented from reaching the image, then information in the shaded areas of the graph is lost from the system. This information, carried on the periphery of the diffraction pattern, is the most important from the point of view of the microscopist, as it corresponds to the *smallest* values of D in the object. The result of cutting off the pattern at θ_{max} is that any features in the object that are smaller than $D_{min} = \lambda/\sin\theta_{max}$ are no longer resolved in the image, because the information about smaller distances has been suppressed.

Even when no restricting aperture exists within the optical system, the objective lens itself limits the maximum angle at which light rays can enter

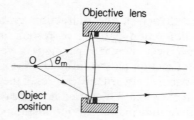

Figure 9.23.   The semi–angular aperture θ_m of a lens

the microscope. The angle θ_{max} is then equal to the 'semi–angular aperture' of the lens, θ_m (Figure 9.23), and the resolution of the microscope is given by

$$D_{min} = \frac{\lambda}{\sin\theta_m}.$$

(9.19)

The wavelength to be inserted in the above equation is that of the light as it leaves the object, but often the microscope specimen is immersed in a medium (e.g. oil) which has a refractive index, μ, much different from that of a vacuum. The vacuum wavelength λ_v of the illuminating light is reduced in the medium to the value $\lambda = \lambda_v/\mu$, so we can write what is a more usual form of equation (9.19), that is

$$D_{min} = \frac{\lambda_v}{\mu\sin\theta_m},$$

(9.20)

where the product, $\mu\sin\theta_m$, is known as the numerical aperture (N.A.) of the objective lens with its immersion medium.

9.4 Polarized Light

As we saw at the beginning of this chapter, light is a transverse oscillation, for the electric and magnetic fields produced by a 'ray' of light are both at right angles to the direction of travel of the waves. In plane–polarized light the vector which represents the electric field always remains in a given plane called the plane of polarization (Figure 9.24). The name is given to

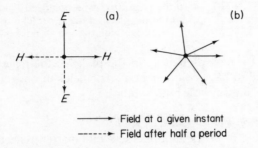

———→ Field at a given instant
------→ Field after half a period

Figure 9.24. View along a light ray showing the field directions. In plane–polarized light (a) the field remains in one plane. Unpolarized light (electric field only is shown) can have fields in any plane (b)

the plane of the electric, rather than the magnetic, field because it is the electric field which is responsible for vision and for the photographic effect. In contrast to plane–polarized light, ordinary light produced by an incandescent source (e.g. a tungsten filament lamp) is unpolarized (i.e. the electric fields, although still transverse to the direction of travel, are otherwise oriented at random and have no common direction).

If one looks along the direction of a ray of light in which the electric field has a particular direction at a particular time (Figure 9.25) the electric

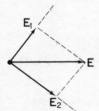

Figure 9.25.   View along a light ray showing resolution of the field into fields which are at right angles. Vectorially,
$$\mathbf{E} = \mathbf{E}_1 + \mathbf{E}_2$$

field vector, $\mathbf{E}$, can be considered as the sum of two other electric fields in chosen directions which are at right angles—the field can be resolved into two perpendicular components. We can thus consider any ray of light, when it is convenient, as the sum of two plane–polarized rays in which the planes of polarization are mutually perpendicular, since the magnitudes of the fields of the 'component' rays can be chosen so as to add correctly to give the field of the original one. Diagrammatically, any ray of light may therefore be shown as in Figure 9.26, where the dots represent an electric field oscillating in a direction perpendicular to the plane of the page and the

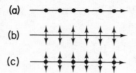

Figure 9.26. Representation of light rays: (a) plane–polarized light with electric field normal to the page; (b) plane–polarized light with electric field in the plane of the page; (c) unpolarized light, or light polarized in an unknown plane

arrows represent a field oscillating in the plane of the page. A general ray may be supposed to have component electric fields in both these directions, and is therefore shown with both symbols.

9.4.1 *Production of Plane–polarized Light*

Most of the ways in which light interacts with matter (reflection, refraction, absorption and scattering) can produce polarized light from unpolarized light. In reflection, when light falls on the surface of a transparent medium, such as glass or water, some of the light is reflected and some is refracted into the medium (Figure 9.27). Obviously some division of amplitude occurs

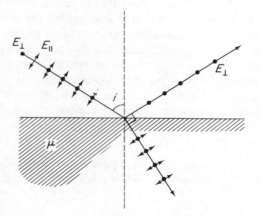

Figure 9.27. Reflection and refraction of light (drawn for the Brewster angle of incidence)

at the surface and it is found that the proportions of the electric field of the incident ray that are reflected or transmitted depend not only on the angle of incidence, i, but also on the orientation of the electric field with respect to the plane of incidence. (The plane of incidence is the plane of the page in Figure 9.27; it contains the incident, reflected and refracted rays.) The percentage of original light intensity which is reflected varies. with i in one way for electric fields in the plane of incidence ($E_\parallel$) and in another way for perpendicular fields ($E_\perp$). This is shown in Figure 9.28 for reflection from a glass plate. With light falling directly on the surface ($i = 0$) only about 4 per cent. is reflected in either case, and at grazing incidence ($i = 90°$)

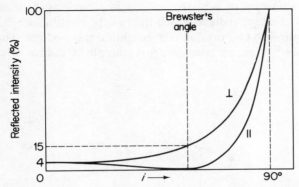

Figure 9.28.   The intensity of light reflected from a smooth glass surface as a function of the angle of incidence. ⊥ is light polarized normal to the plane of incidence and ∥ is light polarized in the plane of incidence

the reflection is 100 per cent. At intermediate angles the reflection curves diverge and at one particular angle (Brewster's angle) the percentage reflection for $E_\parallel$ falls to zero: only light with its electric field at right angles to the plane of incidence ($E_\perp$) is reflected. The reflected light at this angle is thus pure plane–polarized light. The refracted light (the rest) is partially polarized, since it contains all the electric field $E_\parallel$ and the part of $E_\perp$ which is not re-flected, and the ratio of $E_\parallel$ to $E_\perp$ is different from that in the original incident light. In Figure 9.27 the rays are drawn with an angle of incidence equal to Brewster's angle (i_B). At this angle the reflected and refracted rays are at right angles, and it can be shown theoretically that this is the necessary condition for $E_\parallel$ to have zero reflected intensity. For this arrangement of the rays, it follows from the law of refraction that $\tan i_B = \mu$, so that Brewster's angle can be determined if we know the refractive index, μ, of the second medium relative to the first. For glass, with a refractive index of about 1·5 relative to air, Brewster's angle is around 57°.

When light is refracted in a medium, it is found that in some materials there are two refractive indices, one for light with its electric field in one plane and another for the field in another direction. This phenomenon is called double refraction, or *birefringence*. Birefringent crystals, such as calcite and quartz, have different electrical properties along different axes in the crystal, and are said to be electrically *anisotropic*. One particular direction in the crystal is called the *optic axis*. Light with its electric vector perpendicular to this axis travels at one speed in the crystal, while light with its electric vector parallel to the optic axis travels at another speed (usually greater). All rays directed along the optic axis travel with one speed, since all their electric fields, being perpendicular to the ray, are perpendicular to the optic axis, too.

Since two velocities of light are possible in a birefringent material, there are also two refractive indices, which implies the possibility of the incident light being split into two rays on refraction into the medium. This is shown in Figure 9.29, where we consider, for example, a calcite prism with its

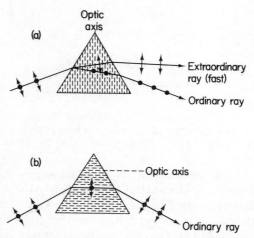

Figure 9.29. Double refraction of light in a birefringent crystal. In (a) the prism is cut with the optic axis vertical and in (b) with the axis horizontal

optic axis (a) vertical and (b) horizontal in the diagram. In the first case unpolarized light is split into rays with two different velocities. The ray with its electric field perpendicular to the optic axis is called the 'ordinary' ray (O), while the one with its field parallel to the optic axis is the 'extra-ordinary' ray (E). In calcite the E ray travels faster than the O ray and therefore has a smaller refractive index and is deviated less. Two plane-polarized rays of light leave the prism. However, in case (b) the unpolarized light entering the prism is not split into two beams at this particular angle of incidence, since for the prism cut from the crystal in this way all the electric fields in the crystal are then perpendicular to the optic axis. In (b), therefore, no separation occurs since all the light has a common velocity and emerges as an 'ordinary' ray.

Calcite is used in the construction of a *Nicol prism*, which is a device for producing a single beam of plane–polarized light. The principle is to split the light rays by birefringence in the way just described, and remove one of the rays completely by total internal reflection at a surface within the prism. The prism is shown in Figure 9.30, and is made from two calcite

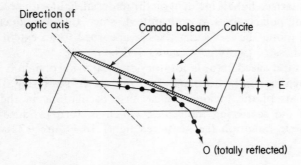

Figure 9.30.   Construction of a Nicol prism

sections cemented together with Canada balsam, a transparent cement with a value of refractive index between those for the O and E rays in the calcite. The advantage of the Nicol prism is that it makes a complete separation between the E ray and O ray, since total internal reflection is an 'all–or–nothing' phenomenon.

A simpler method of producing polarized light is to use a dichroic material, such as Polaroid. Dichroic molecules have two coefficients of absorption, one for one direction of the electric field and one for another direction.

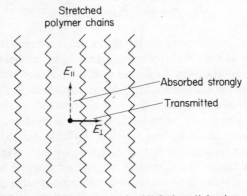

Figure 9.31.   Absorption of plane–polarized light in a dichroic material
such as Polaroid (looking along the ray)

Examples are iodosulphate of quinine and crystals of tourmaline. The synthetic material, Polaroid, was invented in 1932 by Land, who stretched thin films of polyvinylalcohol impregnated with iodine and found that the long–chain polymer molecules were aligned by the stretching. The chains absorb light strongly when the plane of polarization is parallel to the chain axis, but little when the electric field is perpendicular to the axis (Figure 9.31). If unpolarized light is incident on the material, light with predominantly one plane of polarization is transmitted, since the component with its electric field along the molecular axes is absorbed to an extent dependent on the thickness of the film.

Although rarely used in the laboratory, the scattering of light is an interesting phenomenon in which polarization occurs. In scattering, the electro–magnetic fields of the incident light cause oscillations in the electronic structure of the scattering molecules, which in turn produce secondary electromagnetic radiation (i.e. scattered light). In Figure 9.32 we see why

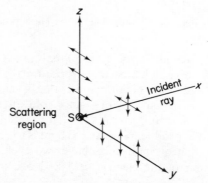

Figure 9.32.   Molecular scattering of unpolarized light. The scattered
radiation is generally polarized

the scattered radiation is polarized when observed in certain directions. Consider the electric field of unpolarized light incident along the x–axis as the sum of oscillations in the y– and z–directions, both normal to the direction of travel. The light strikes scattering molecules at S. First, since there is no electric field *along* the ray, there is no oscillation of electric field in the x–direction which can be re–radiated, so x–vibrations are absent from the scattered light. Secondly, on looking, for instance, along the y–direction, there will be no radiation in this direction of the y–vibration present in the incident light, since this vibration would be along the line of travel, whereas light is only a transverse oscillation. Only z–vibrations of electric field can be radiated in the y–direction and y–vibrations in the z–direction. Thus, the scattered light in Figure 9.32 is vertically polarized when viewed along the y–axis and horizontally polarized when viewed along the z–axis. Since the z– and y–directions may be chosen quite arbitrarily, light scattered at right angles to the incident beam will always be plane–polarized. In directions other than this the light appears partially polarized to an extent depending on the angle of observation.

The scattered blue light from the sky is polarized in this manner; sky light observed in a direction at right angles to the rays of the sun may be up to 80 per cent. polarized, depending on the atmospheric conditions. It is believed that bees can detect the direction of polarization of sky light and are aware, through this, of the position of the sun in the sky even if the sun itself is obscured by cloud. They can successfully navigate between the hive and a source of nectar in this way. Light in the ocean is similarly polarized, to a lesser extent, and various aquatic animals are known to regulate their motion by reference to it.

9.4.2 Detection of Plane–polarized Light

If we use a piece of Polaroid or a Nicol prism as a *polarizer* to produce polarized light from an ordinary light source, the transmitted light has a particular plane of polarization, the orientation of which we know from the geometry of the prism or from the Polaroid, if someone has been good enough to mark a direction on it. It is interesting to see what happens if we now place another Nicol prism or Polaroid in the already polarized beam, and rotate this second optical component (the *analyser*) on the axis of the light beam. The analyser contains a preferred plane in which it will allow an electric field oscillation to be transmitted, that is parallel to the optic axis of the Nicol prism or perpendicular to the polymer axis of the Polaroid. If this plane of transmission of the analyser makes an angle θ with the plane of polarization of the incident light, the situation is as shown in

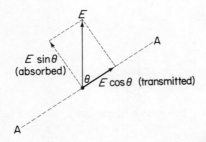

Figure 9.33. Partial transmission of polarized light by an analyser with its plane of transmission AA at an angle θ to the plane of polarization

Figure 9.33. Here we are looking along the direction of the beam and the plane of polarization (containing the electric field) is vertical.

The electric field **E** can be resolved into two component fields at right angles, one of which, in the plane of the analyser, is transmitted, while the other, perpendicular to the plane, is removed from the beam. The trans-mitted value of the electric field is $E \cos \theta$ and the amplitude of the light wave transmitted by the analyser is proportional to this value. However, any light detector such as the eye or a photocell responds to the *intensity* of the light, which is the square of the amplitude. The intensity transmitted is thus proportional to $(E \cos \theta)^2$ and the variation of intensity seen as the analyser rotates is proportional to $\cos^2 \theta$. When $\theta = 0$ or $\theta = 180°$ the analyser transmits maximum intensity; the plane of polarization of the light then coincides with the plane of transmission of the analyser. When $\theta = 90°$ or $\theta = 270°$ the intensity is a minimum, that is when the two planes are perpendicular.

If the plane of transmission of an analyser is known, this gives a method of finding the plane of polarization of an unknown beam of polarized light, since the analyser can be rotated until the two planes are parallel, when the transmitted intensity becomes a maximum. If the beam contains only plane–polarized light, the analyser should give zero intensity on rotation through a further 90°.

9.4.3 *Circularly Polarized Light*

Let us consider what will happen if a ray of plane–polarized light is incident normally on a sheet of birefringent material which has its optic axis parallel to the surface as shown in Figure 9.34; here the plane of pola–

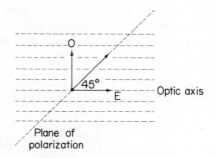

Figure 9.34.   Production of circularly polarized light in a birefringent material (looking along the ray)

rization is at 45° to the optic axis. The figure is drawn looking in the direction of travel of the ray, and we see that the electric field can be resolved into equal components, along the optic axis (E) and at right angles to it (O). The original ray can thus be thought of as the sum of an ordinary ray and an extraordinary ray, polarized at right angles. (Notice that, because the original beam is incident normally on the sheet, it is not deviated on entering, and the E and O waves follow the same path.) Now suppose that in this case the E wave travels faster than the O wave. As the light travels through the material, the E wave gets progressively further ahead of the O wave, so that when the two waves emerge again into air they are out of phase. Since the waves have equal velocities in air, this phase difference then remains subsequently unaltered.

When the thickness of the material is such that the waves have a final phase difference of $\pi/2$ rad, corresponding to a path difference of $\lambda/4$, the sheet is called a 'quarter–wave plate'. The amplitudes of the O and E waves emerging from a quarter–wave plate are shown in Figure 9.35(a), with a

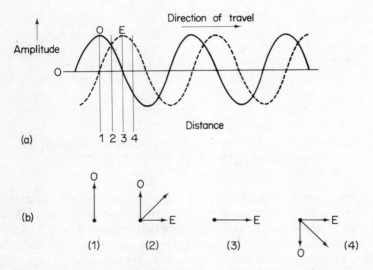

Figure 9.35.   (a) Amplitudes of the O and E waves after emerging from a quarter–wave plate. (b) Combination of the amplitudes of O and E waves (looking along the ray)

phase difference of $\pi/2$ rad or 90°. The figure does not show the *direction* of the electric fields, for of course the E field is perpendicular to the O field in space.

When the O and E waves emerge from the quarter–wave plate we may combine the electric fields of the two, remembering the difference in phase, to find the total electric field in the emergent ray. Figure 9.35(b) shows the effect of recombination for different parts of the wave patterns shown in Figure 9.35(a), again looking along the ray as in Figure 9.34. At point 1 of the wave pattern the O field has a maximum while the E field is zero. The combined electric field is thus vertical (Figure 9.35b). At point 2 the fields of the ordinary and extraordinary ray are equal and the combined field is at 45° to that of point 1. At point 3 the O field is zero and the E field a maximum; the total field is horizontal. At point 4 the O field has reversed in sign while the E field has decreased, and the total field is rotated by 45° from the previous direction.

What the quarter–wave plate has done is to convert the original plane–polarized light into *circularly polarized light*, in which the vector representing the electric field *rotates* around the axis as distance or time increases, instead of oscillating in one plane (Figure 9.36).

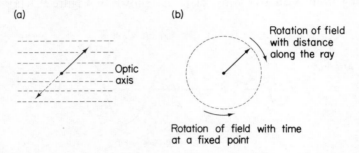

Figure 9.36.   Light (a) before entering a quarter–wave plate (plane–polarized) and (b) after emerging (circularly polarized)

If the plane of polarization of the light incident on the quarter–wave plate were not at 45° to the optic axis, it can be seen from Figure 9.34 that the O and E fields would then have different maximum amplitudes. The result of recombination of the O and E rays emerging from the plate would then be a rotating electric field vector whose tip followed an ellipse rather than a circle, giving *elliptically polarized light*.

Circularly polarized light can obviously have one of two modes of rotation for the electric field vector: clockwise or anticlockwise, looking towards the oncoming light. Some materials, in which the molecules are not symmetrical structures, have the property that one mode of circularly polarized light travels faster in the medium than the other. Such a material possesses one refractive index for each mode of rotation, a phenomenon called circular birefringence but more generally known as *optical activity*. The consequences of passing plane–polarized light through an optically active medium are surprising at first sight, but we can follow what happens, up to a point, by using a mathematical device.

Mathematically, as shown in Figure 9.37, we can regard a ray of plane–polarized light as the sum of two rays of circularly polarized light with

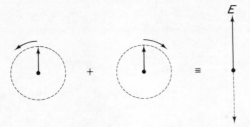

Figure 9.37.   Plane–polarized light as the sum of two circularly
polarized components

opposite modes of rotation and the same frequency as the original light. The amplitude of each rotating vector is half the maximum amplitude of the original electric field. In some plane (the vertical one in the diagram) the two rotating vectors coincide in direction and their sum is a maximum. If we now imagine the vectors to rotate from this position, one to the right and the other to the left, by equal angles, we see that the sum of the two is still in a vertical direction, since the horizontal components of each vector are equal but in opposite directions, so that they cancel out. When both rotating vectors are horizontal their sum is a zero field in any direction and when the two vectors point downwards they coincide to give a maximum field, still in the vertical plane. The addition of the two circularly polarized rays thus gives an electric field oscillating in one plane. (Incidentally, note that the plane of polarization of the total electric field bisects the angle between the two rotating vectors at any time.)

Now consider what happens when plane–polarized light passes through a material with circular birefringence—suppose that the 'clockwise' component of Figure 9.38(a) travels more slowly than the anticlockwise one. The two circular components then have different wavelengths in the material, the slower, clockwise mode having a wavelength, λ_+, which is shorter than that of the anticlockwise mode, λ_-. (See Figure 9.38.) Since the rotating vector makes one rotation in space as we move one wavelength along the ray, the number of complete rotations that occur between the electric field directions on entry to and exit from the material is d/λ if the ray traverses a distance d. Thus, the slower circularly polarized ray, with shorter wavelength, achieves more rotation than the other, the difference in radians $\Delta\theta$ being given by

$$\Delta\theta = 2\pi\left(\frac{d}{\lambda_+} - \frac{d}{\lambda_-}\right). \tag{9.21}$$

When we look at the two rotating vectors after the rays have emerged from the material, at an identical instant and point in space such that the 'anti–clockwise' vector is vertical, we find that the other vector is not vertical

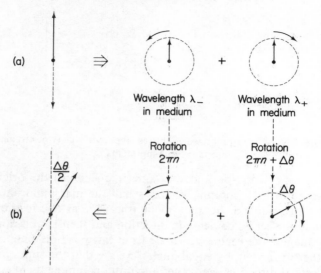

Figure 9.38. The passage of plane–polarized light through an optically active material. (Vectors are drawn as if looking *towards* the light.)
In (a) is the situation on entry of the light and in (b) the vectors are
shown after the anticlockwise vector has made a complete number
of rotations

(Figure 9.38b) although originally the two vectors coincided in the vertical
plane. The clockwise vector is rotated by the extra angle $\Delta\theta$ introduced by
the birefringence of the optically active material. When we now recombine
the electric fields of the two circular components, they still give a plane–
polarized electric field, but the direction, which bisects the angle between
the component vectors, is now rotated $\Delta\theta/2$ to the right.

The effect of an optically active material is therefore to *rotate* the plane
of polarization of light which passes through it by an amount α given by

$$\alpha = \frac{\Delta\theta}{2} = \pi d \left(\frac{1}{\lambda_+} - \frac{1}{\lambda_-} \right). \tag{9.22}$$

The amount increases with the path length, d, in the material, and for small
velocity and wavelength differences ($\lambda_+ \simeq \lambda_-$, $\lambda_- - \lambda_+ = \Delta\lambda$) we can write

$$\alpha = \pi d \frac{\Delta\lambda}{\lambda^2}. \tag{9.23}$$

This inverse–square dependence on wavelength is known as Biot's law,
and we see that α should decrease on going towards the red end of the
spectrum. The variation of α with wavelength is known as 'optical rotatory
dispersion' (O.R.D.) and when in some cases it does not obey Biot's law
the dispersion is said to be anomalous.

When α is positive (rotation of plane–polarized light to the right when looking against the light) the material is said to be dextrorotatory and when negative (rotation to the left) is said to be laevorotatory. A common substance that is laevorotatory is turpentine, while the sugar dextrose, as its name implies, is dextrorotatory. Some crystalline materials, such as quartz, have a crystal 'habit', or growth form, which may be either left–handed or right–handed, one form being laevorotatory while the other is dextrorotatory.

9.4.4 Polarized Light in the Study of Molecules

Many biologically important molecules are optically active and rotate the plane of polarized light which is passed through a solution of them. Since in liquids the amount of rotation in a given path depends on the concentration of the molecules present, the measurement of optical rotation for a known substance is a means of determining its concentration in solution; this is a standard method for sugars.

The instrument used to measure optical rotation is called a *polarimeter* (sometimes a *saccharimeter* when used specifically for sugars) and is shown schematically in Figure 9.39. A beam of light from a monochromatic source

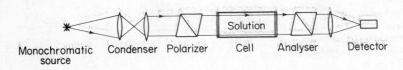

| Monochromatic source | Condenser | Polarizer | Cell | Analyser | Detector |

Figure 9.39. Schematic diagram of a single–wavelength polarimeter

(e.g. a sodium lamp) passes through a polarizer, and plane–polarized light then emerges to enter a cell with transparent end windows, which contains the solution. The path length in the solution must be known accurately and the cell is often maintained at a fixed temperature by circulating water around it. The light emerging from the cell passes through an analyser and is focused either on a photoelectric device or into some optical device which enables the eye to detect easily a minimum light intensity. The analyser is rotated so that it transmits a minimum light intensity, and its plane of transmission is then perpendicular to the plane of polarized light leaving the cell. This is done first with a pure solvent in the cell and then with the solution, and the difference in the settings of the analyser gives the optical rotation, α, produced by the solute molecules alone.

In dilute solutions

$$\alpha = [\alpha]\, l\, c, \tag{9.24}$$

where $[\alpha]$ is a property of the substance alone, the specific rotation (or specific optical rotatory power), and l and c are respectively the path length and concentration in the cell. In the formerly accepted definition of $[\alpha]$, α is in degrees, l in decimetres and c in gm cm^{-3}. In the S.I. definition, α is in radians, l in metres and c in kg m^{-3}. Since α depends on wavelength and temperature, $[\alpha]$ is written with both values (e.g. $[\alpha]_D^{20}$ is the specific rotation at 20°C for the sodium D–line in the spectrum). In more complex polarimeters the wavelength of the light used can be varied to obtain a graph of $[\alpha]$ against λ.

Table 9.1 contains values of $[\alpha]$ for a number of molecules of biological interest. Most amino acids and proteins are optically active and it is possible

TABLE 9.1

Values of the specific optical rotatory power (specific rotation)
for a number of biological substances in aqueous solution

Solute	$[\alpha]_D^{20}$ (deg dm^{-1} cm^3 gm^{-1})	$[\alpha]_D^{20} \times 10^4$ (rad m^2 kg^{-1})
Alanine	+1·8	+3·14
Alanine (in HCl)	+14·5	+25·3
Leucine	−11·0	−19·2
Cysteine	−16·5	−28·8
Glutamic acid	+12·0	+20·9
Bovine serum albumen	−63	−110
Bovine serum γ–globulin	−46	−80
β–casein	−112	−195
Native collagen	−280 to −400	−489 to −698
Denatured collagen	−110 to −135	−192 to −236
Insulin	−34	−59
Myosin	−28	−49
Pepsin	−69	−120
Trypsin	−40	−70

A positive sign indicates a dextrorotatory substance, while a negative sign indicates a laevo–rotatory one.

to infer something about their molecular structure from the sign of the rotation which is produced. For instance, the alanine molecule has two possible asymmetric structures, one the mirror image of the other. One (L–alanine) rotates polarized light anticlockwise and the other (D–alanine) rotates it in the opposite sense. In fact, it is found that all naturally occurring amino acids and proteins are in the L–form and have a negative specific rotation. Helical protein molecules have large negative values of $[\alpha]$, and when such a protein changes to a different shape (e.g. at a different pH or temperature) there is a consequent change in $[\alpha]$, so that these changes in shape can be easily detected and measured and the properties of the molecule studied under various conditions. In addition, the optical rotatory dispersion of helical proteins is anomalous at short wavelengths, and by studying how $[\alpha]$ varies with λ in the ultraviolet region, it is possible to determine how much of a polypeptide chain is twisted into a helical form and thus to estimate the 'percentage helix' present in globular proteins such as albumin.

Further Reading

Jenkins, F.A., and H.E. White (1957). *Fundamentals of Optics*, 3rd ed. McGraw–Hill, New York.

Jirgensons, B. (1969). *Optical Rotatory Dispersion of Proteins and Other Macromolecules.* Springer–Verlag, Berlin.

Kingslake, R. (Ed.) (1965). *Applied Optics and Optical Engineering*, Vol. 1. Academic Press, New York.

Webb, R.H. (1969). *Elementary Wave Optics.* Academic Press, New York.

PROBLEMS

9.1 Two trains of plane light waves are travelling in the same direction, and the maxima of one train occur an eighth of a wavelength behind those of the other, which are of the same amplitude. What is the phase difference between the two wave trains? At a point where the electric field strength of one of the waves is a maximum, E_0, what is the electric field strength in the other wave train?

9.2 An amoeba is viewed in an interference microscope under conditions in which the cell is effectively between the two surfaces of a wedge–shaped film of water. The parallel interference fringes which are seen in the film are shifted to one side where they cross the image of the cell. The shift caused by a particular region of the cell is equal to half the distance between one fringe and the next. If the wavelength of the light used is 546 nm and the region has a thickness of 5 μm, calculate the difference between the refractive index of the cell contents and that of water.

9.3 A diffraction grating with 10^4 lines per inch is illuminated with mono–chromatic plane–parallel light incident along the normal to the surface. The second–order diffracted maximum is observed at an angle of 27° 40′ from the incident beam. Calculate the wavelength of the light. How many orders of diffraction would you expect to be able to see? If the third order were missing, what information could be inferred about the grating?

9.4 According to Babinet's principle, the set of haloes seen when viewing a distant light source through a suspension of identical spherical particles has the same form as the Airy pattern from an aperture of the same diameter as the particles.

In the haloes seen round the moon on a misty night the innermost dark ring appears to have a diameter twice that of the moon's disc. (The moon has a diameter of 2 160 miles and is about 240 000 miles from the earth.) Find the average diameter of the water droplets in the mist, assuming that the average wavelength of moonlight is 450 nm.

9.5 A certain microscope objective lens accepts light scattered by the specimen at angles up to $60°$ from the directly transmitted light beam and is used with an immersion oil of refractive index 1·50. What is the numerical aperture of the objective? If the specimen is illuminated by a parallel light beam which has a wavelength *in vacuo* of 546 nm, estimate the resolution of the microscope in these circumstances.

9.6 A certain birefringent crystal has refractive indices of 1·5442 and 1·5563 for the E and O waves respectively, at a wavelength of 600 nm. Calculate the least thickness of crystal necessary to produce an optical path difference of a quarter of a wavelength between the E and O waves which pass through it.

If polarized light is passed through such a quarter–wave plate and the plane of polarization of the incident light lies at $30°$ to the direction of the optic axis, what kind of polarized light will emerge from the plate? If the emerging light is observed through an analyser which is rotated through $180°$, what ratio will exist between the maximum and minimum observed intensities?

9.7 The optical rotation of plane–polarized light in passing through 100 mm of a sucrose solution is $11·6 \times 10^{-3}$ rad (c.0·7°) when the concentration of the sucrose is 10 kg m^{-3} and the wavelength of the light is 589 nm. Use Biot's law to estimate the fractional difference in wavelength $(\Delta\lambda/\lambda)$ of right–handed and left–handed circularly polarized light passing through the solution. What would you expect the optical rotation to be at a wavelength of 196 nm?

CHAPTER 10

Optical Microscopy

The ultimate aim of a light microscope is to magnify an object so that details invisible to the naked eye may then be observed. With a microscope individual cells of, for example, a leaf can be seen while small organisms such as protozoa and bacteria, which are normally invisible, are observed moving, apparently rapidly, through their fluid environment.

A *simple* microscope consists of a single convex lens which produces a magnified image, but which can produce only a limited amount of magnification before aberrations seriously impair the image quality (Chapter 8).

10.1 The Compound Microscope

The microscope as we know it is a compound one in which several lenses are used. In the compound microscope the required magnification is achieved in two stages (Figure 10.1). An *objective* lens produces an intermediate image,

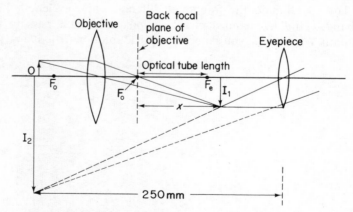

Figure 10.1.   Construction of a compound microscope

I_1, which is further magnified by the eyepiece thus producing a final virtual image, I_2, which is seen by the observer. The image may be formed at the least distance of distinct vision (conventionally 250 mm), in which case equations (8.13) and (8.15) of Chapter 8 can be used to show that the overall magnification, M, of the microscope is given by

$$M = \frac{x}{f_o}\left(\frac{25}{f_e}+1\right).$$ (10.1)

In this equation the expression in brackets represents the magnification due to the eyepiece, which has a focal length f_e, while x/f_o is the magnification produced by the objective, focal length f_o. The distance x is that between the intermediate image and the second focal point of the objective lens. In producing a large overall magnification, the objective usually plays a larger part than the eyepiece. From equation (10.1) it can be seen that a high–power objective must have a short focal length, f_o. For example, an objective producing an intermediate image ten times as big as the object in a standard microscope has a focal length of 16 mm, while for a magnification of forty times the focal length is 4 mm.

When large overall magnifications are used or if the microscope is used for long periods, an observer would quickly become tired through eyestrain by observing an image at 250 mm. It is much more restful to view the final image at infinity, so that the eye is completely relaxed during observation. This reduces the magnification a little (the bracket in equation (10.1) becomes $25/f_e$), but this disadvantage is more than offset by the reduction in the fatigue suffered by the operator and can be compensated by selecting appropriate combinations of objective and eyepiece to make up the desired magnification. For a particular microscope a range of objectives is normally available with magnifications varying from $\times 2$ to $\times 100$ (read $\times$ as 'times'), while eyepieces with a magnification range of $\times 5$ to $\times 25$ may be obtained. This means that the maximum overall magnification of the microscope is $\times 2500$. Lenses could be made to increase this magnification still further but, as will be shown later, no further detail is made visible by the increased magnification.

10.2 Objectives

The lenses of a real microscope are more complicated than the single biconvex ones shown in Figure 10.1 because of lens aberrations (Chapter 8). Examples of the complex arrangements of lenses necessary to reduce the aberrations in an objective were shown in Figure 8.18. The arrangement of lenses is cemented inside a metal cylinder which is threaded at one end so that it can be screwed into the tube of the microscope (Figure 10.2). Because the objective is intended to operate at fixed object and image distances, its magnifying power is known. This figure is usually inscribed on the barrel of the objective together with the focal length and numerical aperture, a parameter which will be discussed in Section 10.4.1. In addition, objectives are sometimes colour–coded so that their magnifying powers

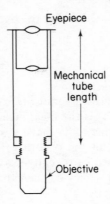

Eyepiece

Mechanical
tube
length

Objective

Figure 10.2. Mechanical details of the structure
of a compound microscope

can be recognized quickly. Thus a green band indicates a $\times 16$ objective
while a $\times 100$ lens has a white band.

Specimens on a microscope slide are normally under a cover slip. This
thin piece of glass will modify the paths of light rays from the specimen to
the microscope and is considered when the objective is designed. It is
possible to correct the lens aberrations for only *one* thickness of cover slip,
which must be used if the optimum performance is to be obtained from the
instrument. Objectives are usually designed for a cover slip thickness of
from 0·17 mm to 0·18 mm, generally obtainable as number $1\frac{1}{2}$ cover glasses.

10.3 Eyepieces

Eyepieces, or *oculars*, also usually consist of more than one lens, although
they are not as complex as the objectives in their construction. The lenses
are again mounted in a cylinder which simply slips into the microscope
tube (Figure 10.2) and is held in position by its shoulder.

10.3.1 *Huygens Eyepiece*

The Huygens eyepiece (Figure 10.3a) is the most frequently used type
and consists of two planoconvex lenses separated by a distance equal to
the mean value of their focal lengths. The image, I_1, formed by the microscope
objective occurs between the field lens and the eye lens and acts as a virtual
object for the field lens. The focal length of the field lens is two or three
times that of the eye lens. A real image is produced at the focal point of the
eye lens so that the final image can be viewed at infinity. The field stop in
the eyepiece is placed at the focus of the eye lens so that it is sharply focused
with the final image. The boundary of the field observed will thus be clearly
defined and will not have a distracting fuzzy edge. It is often useful to place
a graticule or cross wires in the eyepiece for measuring or for other purposes.
Since this graticule must be sharply focused at the same time as the image,
it is placed in the same plane as the field stop.

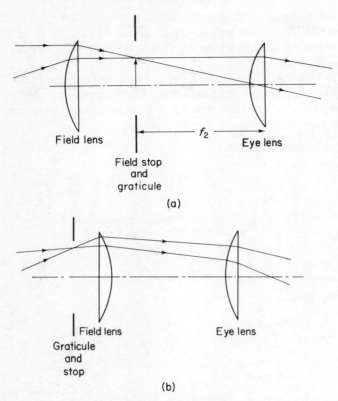

Figure 10.3.   Types of eyepiece: (a) Huygens, f_2 = focal length of eye lens, and (b) Ramsden

10.3.2 *Ramsden Eyepiece*

In the Ramsden eyepiece (Figure 10.3b) the focal plane of the eye lens, and hence the position of a graticule, is outside the lens system. This is a useful feature when measurements are made by moving the graticule across the image by means of a micrometer. The Ramsden system consists of two planoconvex lenses of about the same focal length separated by a distance of about three–quarters of this focal length. The latter condition means that any dust particles on the field lens will not be sharply in focus.

10.3.3 *Other Eyepieces*

Other types of eyepiece are made for special purposes. For example, *compensating* eyepieces (which are marked with the letter K) are designed to compensate for the chromatic difference in magnification (lateral colour, Chapter 8) which gives coloured fringes round the image and occurs for high–power objectives, particularly of the apochromatic type. *Projection* eyepieces, used to display an image on a screen or in some cases for photo-graphic work, are usually of the Huygens type with an adjustable eye lens for focusing purposes.

Modern research microscopes have a binocular arrangement so that both eyes may be used to view the final image. Microscopes equipped in this way are far less tiring to use than those with a single ocular. An arrangement of

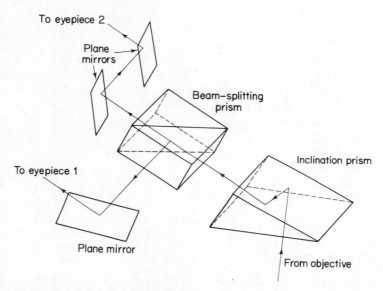

Figure 10.4.   An optical arrangement which permits binocular viewing of a microscopic specimen. A single beam is divided at the beam-splitting prism and reflected by mirrors (or totally reflecting prisms) in such a way that the optical path length to each eyepiece is the same. The inclination prism allows the eyepieces to be set at an angle to the microscope, thereby creating more comfortable viewing conditions for an observer

prisms is used (Figure 10.4) so that the light paths to the two eyepieces are of the same length, thereby permitting both eyes at once to focus on the image.

10.3.4 *Tube Length*

We have already noted that microscope objectives are designed to be used at a single object (and image) distance. Since there is no provision on modern microscopes for adjusting the height of the eyepiece in a particular microscope, the position of the images produced by each objective in turn must fall in the same plane if the final image is to be sharply focused. Different manufacturers sometimes use different image distances, so it is recommended that objectives are used only on the type of microscope for which they were designed.

The term 'tube length' is often encountered in literature concerning microscopes. There are, in fact, two different meanings of the term. One, the *optical* tube length, is the separation of the foci of the eyepiece and objective (i.e. F_oF_e in Figure 10.1). The other, the *mechanical* tube length, is the distance between the objective screw–in surface and the eyepiece rest surface (Figure 10.2); this distance is 160 mm for the majority of modern microscopes.

10.4 Resolution

The ability of a modern optical microscope to render visible details in an object or to separate two closely spaced object points is governed by the nature of light itself rather than by imperfections in the lenses. We say that the microscope is a *diffraction–limited* system. In Chapter 9 it was shown that, for an object illuminated by parallel light and viewed through a lens, the smallest resolvable distance, D_{min}, in the object is

$$D_{min} = \frac{\lambda}{n \sin \theta}, \tag{10.2}$$

where λ is the wavelength of light used, n is the refractive index of the medium between the object and the lens, while θ is half the angle subtended by the lens at the object. In a microscope operating under optimum conditions a suitable arrangement of lenses below the specimen provides a cone of light

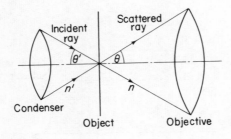

Figure 10.5.   Paths of extreme rays from the condenser to the objective of a microscope. n^1 and n are the refractive indices of the media *between* the two lenses and the object

(Section 10.5). From Figure 10.5 it is clear that light can be collected by the lens at a greater angle of scattering in this case than is possible when light parallel to the axis of the system is used to illuminate the specimen. It can be shown that the smallest resolvable distance under these conditions is

$$D_{min} = \frac{\lambda}{n^1 \sin \theta^1 + n \sin \theta}.$$ (10.3)

When the microscope is correctly adjusted

$$n^1 \sin \theta^1 = n \sin \theta.$$ (10.4)

10.4.1 *Numerical Aperture*

The quantity $n \sin \theta$ is called the *numerical aperture* (N.A.) of the objective. Thus equation (10.3) may be rewritten as

$$D_{min} = \frac{\lambda}{2\text{N.A.}}.$$ (10.5)

The numerical aperture governs both the resolving power of the lens and the ability of the lens to collect light. (The first property depends heavily upon the second.) It is similar in the second respect to the *f*–number of a camera lens, and Table 10.1 shows equivalent values of the two quantities.

TABLE 10.1
Equivalent numerical apertures and *f*–numbers

N.A.	0·1	0·25	0·50	1·0	1·4
f–number	5·0	2·0	1·0	0·5	0·36

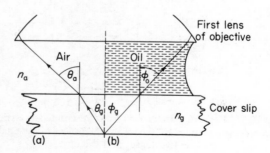

Figure 10.6. Paths of extreme rays collected by the objective, showing the effect of a cover slip and the medium between the cover slip and objective. In (a) the medium is air while in (b) it is an oil having a refractive index equal to that of the glass

Figure 10.6(a) shows how rays from the object pass through the cover slip before reaching the objective lens. The N.A. of the objective is $n_g \sin \theta_g$ or $n_a \sin \theta_a$ (the two quantities are the same because of the laws of refraction). The light–collecting power is clearly improved if immersion oil is used between the cover slip and the objective (Figure 10.6b), since the oil has practically the same refractive index as the glass and hence no refraction occurs as light emerges from the upper surface of the cover slip. The cone of light in the cover slip which will eventually be collected by the objective has a larger angle (ϕ_g) at its apex when immersion oil is used than when air is the medium beneath the objective.

10.4.2 Improvements in Resolution

From the foregoing analysis it can be seen that the resolving power of the objective depends upon three quantities: (a) the wavelength of the light used to illuminate the object, (b) the angular aperture, θ, of the objective and (c) the refractive index of the medium between the objective and the cover slip.

By reducing the wavelength of the light, the resolving power can be improved up to a point. Often green or blue light is used for this reason, but in many cases this is of no advantage since the coloured image produced by white light allows the identification of small details which would remain undetected in monochromatic light. The lowest wavelength which can be used for visual observation is about 400 nm, but use can be made of ultra–violet light if photographic or photoelectric recording devices are used. To gain much advantage from these wavelengths, however, special quartz optics must be used since most types of glass will not transmit radiation of wavelength less than about 360 nm (Chapter 7).

The resolving power of the microscope can also be improved by increasing the angular aperture, so that the value of $\sin \theta$ increases and the value of D consequently decreases. The maximum value for $\sin \theta$ is unity, corresponding to an angular aperture of 90°, but this obviously cannot be achieved in practice as the objective does not touch the object. It is possible to obtain an angular aperture of 71° with real objectives; this corresponds to $\sin \theta = 0.95$, so little improvement in resolution can be expected by further increasing the value of θ.

Increased resolving power can also be attained by putting liquids (immersion fluids) between the objective and specimen so that the value for the refractive index is made larger than unity (which corresponds to air). Water is sometimes used as an immersion fluid, especially in some applications where no cover slip is used. The objective is then in contact with the medium containing the specimen. It is, of course, necessary to take account of the *absence* of the cover slip when the objective is being designed. Water has a refractive index of 1.33, but better results can be obtained with immersion oils which have refractive indices of about 1·5 and do not tend to evaporate as water does. Oils must, of course, be used in conjunction with a cover glass and they must be quite transparent.

The techniques available to the designers and the manufacturers of modern microscopes have resulted in instruments which are capable of achieving almost the maximum theoretically possible resolution. The criteria for resolving power were derived for a special type of object (Chapter 9) consisting of alternate transparent and opaque regions. Such objects are rarely, if ever, encountered in biological materials, which means that, in practice, the limit of resolution often falls short of that which is theoretically attainable. Special techniques (e.g. staining, phase–contrast microscopy) have been devised to improve the contrast between an object and its surroundings, but in some cases artifacts are introduced into the image and the resolution often suffers. When using these techniques great care has to be exercised in the interpretation of the image.

10.4.3 Useful Magnification

We have all noticed at one time or another that when we sit too close to a television screen, the picture, although it is larger, is no clearer than when we are more distant from the screen. This is because the resolution in the picture is fixed both by the scanning lines which make up the picture and by the granularity in the screen itself. When we sit close to the screen the resolved detail in the picture is bigger than the detail which the eye can resolve, so that the picture is not improved. This is an example of *empty magnification.*

There is a limit to the useful magnification in a microscope, beyond which the quality of the image does not improve. This limit is reached when the resolution of the image is the same as that of the eye; any efforts to increase the magnification beyond this limit are wasted, as far as the resolution is concerned.

Let us work out the maximum useful magnification for a microscope, assuming that the final image is formed at 250 mm from the eye. Suppose the smallest detail visible in the specimen is given by equation (10.5) so that, taking an average value of 550 nm for the wavelength of light, the resolution is (275/N.A.) nm, where N.A. is the numerical aperture of the objective. Since the eye can just resolve two points separated by 70 μm at a distance of 250 mm, the magnification necessary to make the detail in the microscope specimen visible to the eye is 70 000 × (N.A./275), or about 260 N.A. It is, in fact, uncomfortable to work at the limit of resolution of the eye. It is therefore conventional to use a figure nearly four times greater than that given and to quote the useful magnification of the microscope as 1000 N.A. For example, if an objective with a numerical aperture of 0·65 is used, the useful magnification of the microscope is 650. If the objective has a magnification of × 40, the eyepiece to use would be one with a magnification of × 15, yielding an overall magnification of × 600.

10.5 Illumination of the Object

With all but self–luminous objects an external source of light is required to illuminate the object. Correct use of the illumination is essential if the best possible performance is to be derived from the instrument. It is important that the angular aperture of rays passing through the object point be under the control of the microscopist and be sufficient to fill the whole aperture of the objective. Further, the illumination should be restricted to that area of the object which is to be observed, so that light is not scattered into the objective from regions outside the field of view. Such scattered light produces a background haze and reduces the contrast in the image. It is also important that the field of view be uniformly illuminated.

Two types of source are used in microscopy: uniformly diffusing sources of large extent (e.g. a north sky or an opal bulb) and small, bright sources such as high pressure arc lamps or certain coiled filament lamps. Adequate illumination for low–power objectives can often be provided by means of an extended diffuse source and a plane or concave mirror, together with an iris diaphragm below the specimen to limit the angular aperture of the rays. It is usually more convenient, however, to use a system of lenses and apertures below the specimen to control the illumination. Such a system is called a *sub–stage condenser*, and is essential when high–power objectives are used, to ensure that the aperture of the objective is filled with light.

10.5.1 *Critical Illumination*

One common way in which a sub–stage condenser is used with a large source is to form an image of the source in the same plane as the object (Figure 10.7), a technique known as *critical illumination*.

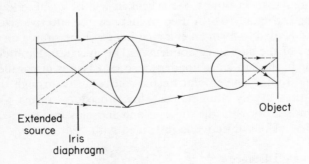

Figure 10.7. Critical illumination with an Abbe condenser

The Abbe condenser illustrated in Figure 10.7 is a commonly used type, but suffers from spherical and chromatic aberrations, so that the image of the source is imperfect. Condensers corrected for these defects are available and are used to obtain maximum performance. The numerical aperture of an Abbe condenser approaches unity, so that it can be used with all 'dry' objective lenses; when used with immersion oil between the slide and the top lens of the condenser a numerical aperture of about 1·4 is possible.

10.5.2 Köhler Illumination

If the light source is small it is difficult to provide even illumination over the entire field of view. In this case Köhler illumination is used (Figure 10.8). Each point on the source yields a parallel pencil of light passing through

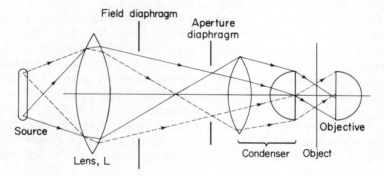

Figure 10.8.   Arrangement of lenses and stops used to achieve Köhler illumination

the specimen. This has the effect that, even if the source itself is brighter in some regions than in others, the specimen will be evenly illuminated.

The lens L (Figure 10.8) is used to increase the effective size of the light source and is used to form an enlarged image of the source on the aperture diaphragm of the sub–stage condenser. The aperture diaphragm is at the focal plane of the sub–stage condenser so that beams of parallel light (each originating from a single point on the source) pass through the specimen, which is thus evenly illuminated. This procedure means that the field diaphragm is in focus in the object plane and may be adjusted so that the field of view of the objective is just filled with light.

10.6 Working Distance

Since the term 'working distance' is often confused with the focal length of the objective, a brief explanation of the term will be given here. The working distance of an objective is the distance between the objective and the specimen, when the latter is sharply focused. If the objective were a simple thin lens, then equation (8.15) could be used to calculate the working distance, since the image distance is fixed. It would be found that the working distance is greater than the focal length of the objective. In practice, objectives are complex arrays of lenses, and their behaviour is not expressed accurately by equation (8.15). The desire to obtain maximum resolution affects the working distance, since it is necessary for the front lens component of the objective to subtend as large an angle as possible at the specimen (equation 10.2). Clearly, this angle increases as the separation of lens and specimen is decreased. It turns out that the working distance of most objectives is much less than their focal length. For example, a ×40 objective with a focal length of 4 mm may have a working distance of only 1 mm to give a numerical aperture of 0·7. Again, an oil immersion objective of N.A. 1·3 with a focal length of 2 mm and a magnifying power of ×100 has a working distance of only 0·12 mm.

In certain circumstances, long–working–distance objectives are desirable, particularly if manipulation of a specimen is to be carried out while it is on the microscope stage. Such objectives can be designed, but only at the expense of the resolution.

10.7 Special Optical Techniques

If the image formed in a microscope is to be seen by an observer, contrast must exist between the image and its surroundings. In many cases this contrast results because the object absorbs light to a greater or smaller extent than the surroundings. If the object absorbs more than the background and monochromatic radiation is used to illuminate the object, the light contributing to the image itself has a smaller amplitude than light reaching the surroundings, so that the image appears dark on a light field (Figure 10.9a). In the case of an object absorbing less than the background the image will appear light on a dark field. In white light the object may appear coloured, according to its pigmentation.

Many biological materials are transparent and are often suspended in a transparent medium. An analogous situation would be glass beads in a vessel of water; the glass beads are very difficult to detect visually. In a microscope using the system described in the earlier part of this chapter, often called a bright–field microscope, such objects would be invisible. This type of object can, however, be rendered visible by making use of the difference in refractive index which usually exists between the object and its environment. This is done in dark–field, phase–contrast and interference microscopes. The type of object described here is often called a 'phase object' to distinguish it from an 'amplitude object' which absorbs light. Because it has a different refractive index from the surroundings, a phase object will scatter light and also retard the light rays passing through it relative to light rays travelling through the surroundings (Figure 10.9b). The retardation changes the relative phase of light emerging from the object (Chapter 9). To see a phase object the scattered light can be viewed by itself, as in a dark–

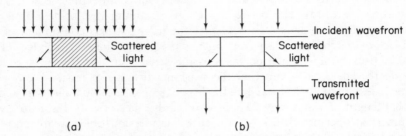

Figure 10.9. (a) Contrast by absorption. The object absorbs more light (as shown by the number of arrows) than its surroundings, and will therefore appear dark against a light background. (b) shows a phase object. Here the portion of the wavefront which passes through the object travels more slowly than the portion traversing the surroundings. The amplitude of the transmitted light is the same for the light which passes through the object and the surroundings, so that the object will remain invisible unless special methods are used to convert the distortion of the wavefront into an amplitude difference

field microscope, or the phase changes introduced by the object can be transformed into amplitude changes in the image, as in phase–contrast and interference microscopy.

Effects resulting from the wave nature of light (Chapter 9) are used in interference and polarizing microscopes to obtain quantitative information about specimens. For example, specimen thickness can often be determined with an interference microscope, while polarizing microscopes can be used to obtain information about birefringence (Chapter 9) in certain types of specimen.

Further information concerning the types and operation of the microscopes described below will be found in the references at the end of the chapter. The brief discussion presented here is intended to introduce the reader to the modern instruments which are available and to indicate in a general manner the type of instrument to be used for a particular purpose.

10.7.1 The Dark–field Microscope

The difference between a bright–field and a dark–field microscope lies in the sub–stage condenser. A dark–field condenser is constructed so that the specimen is illuminated by a hollow cone of light, such that, if the specimen were removed, no light would enter the objective lens (Figure 10.10). The specimen causes light to be scattered into the objective lens and is thereby made visible. The scattering occurs from regions where there is a sharp change in the refractive index, a situation which usually applies at the boundary of a specimen. Such boundaries will appear brightly illuminated

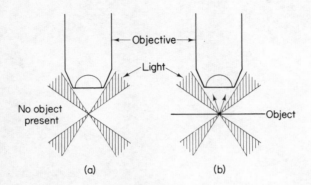

Figure 10.10.   Showing the hollow cone of light used to achieve dark–field illumination. In (a) no object is present and so no light enters the objective, while in (b) the object scatters light into the objective

while a region of uniform refractive index (e.g. the surrounding field) will appear black, producing an image of the type shown in Figure 10.11. Great care must be taken in the interpretation of dark–field images, since spurious interference effects can introduce confusing artifacts.

For high–power objectives the hollow cone of illumination must have a large semi–angle if the light is not to enter these objectives of large numerical aperture. In fact, it is difficult to produce the necessary angle by means of a refracting system, so most dark–field condensers are of the reflecting type. One common type is the paraboloid condenser, illustrated in Figure 10.12. To maintain the large angular aperture available from the reflecting system it is common to use immersion oil between the upper component of the condenser and the lower surface of the slide, thereby eliminating refraction at glass–air boundaries.

10.7.2 *The Phase–contrast Microscope*

Before considering the operation of a phase–contrast microscope, we will discuss in a little more detail how an image is formed in a bright–field instrument. We have already noted that an object diffracts or scatters the light incident upon it; let us consider the special case of a diffraction grating

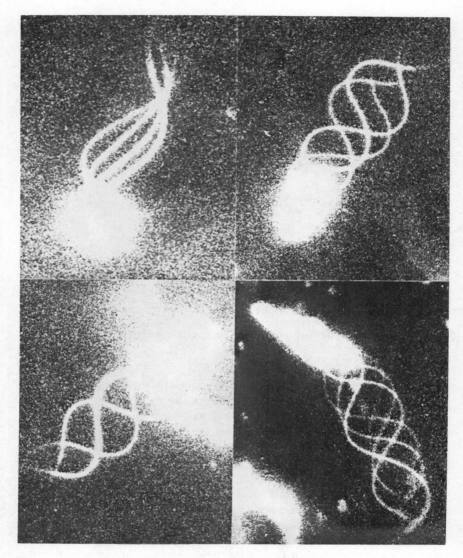

Figure 10.11. Multiple–flash micrographs of flagellated microorganisms taken under dark–field conditions

acting as an object with the microscope condenser adjusted for Köhler
illumination. It follows from the theory of Chapter 9 that diffracted rays
will occur at angles determined by the wavelength of the radiation used to
illuminate the grating, and also by the spacing of the grating. The objective

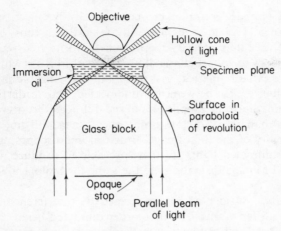

Figure 10.12. Path of light through a paraboloid condenser

of the microscope collects the diffracted rays (Figure 10.13) and focuses them at the back focal plane of the objective. The image of the grating is formed not at this plane but in a plane, I, further from the objective, where the diffracted rays overlap and interfere with each other. The back focal plane of the objective contains the diffraction pattern of the object, a fact

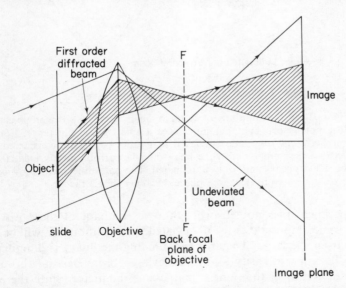

Figure 10.13. Image formation by interference between the undeviated and first-order diffracted beams

which is used in the electron microscope to observe electron diffraction patterns (Chapter 11). The 'true' image of the grating is thus formed by the interference of rays diffracted by the object. To form a correct image it is necessary in theory to collect all the diffracted rays, which is why microscope objectives have large numerical apertures, but recognizable images can be produced by interference between the undeviated (zero–order) beam and a first–order beam. Since the amplitudes of the diffracted beams of high order are much less than those of low order, the former will not significantly affect the visibility of the final image. Practical microscopes, which do not collect all the diffracted light, are therefore able to produce good images. The principles of image formation are the same for biological specimens as for the grating.

For simplicity, consider an image formed by interference between the zero– and first–order beams. The first–order diffracted beam can be shown to be about $\pi/2$ rad out of phase with the undeviated beam (Figure 10.14).

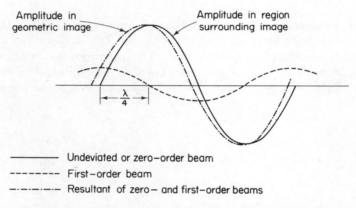

Figure 10.14. Amplitudes of waves in the image space when an object does not absorb light but introduces a small phase difference. For simplicity only the undeviated and first–order beams are considered. Because the amplitudes in the image and its surroundings are equal, the field appears uniformly illuminated. The path difference of $\lambda/4$ is equivalent to a phase difference of $\pi/2$ rad

If the specimen does not absorb light, the resultant of these waves will have the same amplitude as the undeviated beam, although it will be shifted in phase from the latter. The area surrounding the image is illuminated by light of the same intensity as the undeviated beam passing through the object, and is thus of the same brightness as the image. Since the eye can only detect differences in brightness and not in phase, the image will be invisible, as noted earlier.

In the phase microscope the undeviated beam is retarded by $\pi/2$ to produce the situation shown in Figure 10.15(a). The undeviated and first–order beams are now in phase, so that the resultant (which illuminates the image area) is greater in amplitude than the undeviated beam (which illuminates

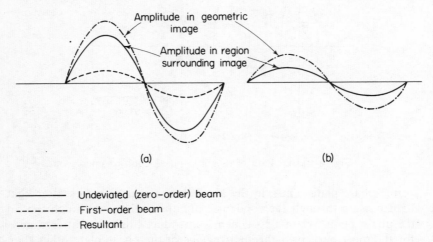

Amplitude in geometric image

Amplitude in region surrounding image

(a)   (b)

——————— Undeviated (zero–order) beam

— — — — — First–order beam

—·—·—·—·— Resultant

Figure 10.15. Relationships between undeviated and first–order diffracted beams when (a) the undeviated beam is retarded by $\pi/2$ rad and (b) the undeviated beam is retarded by $\pi/2$ rad and reduced in amplitude to match that of the first–order beam. In (b), therefore, the solid line represents both the zero– and first–order diffracted beams. An amplitude (and hence intensity) difference exists between the image and its surroundings, so that the image can be seen

the surroundings). Greater contrast is achieved by reducing the intensity of the undeviated beam so that it matches that of the diffracted beam (Figure 10.15b). Since the amplitude of the resultant of the two is now twice the amplitude of the undeviated beam, the image will appear four times as bright as its surroundings, an obvious improvement over the situation shown in Figure 10.15(a).

These conditions are achieved in the phase–contrast microscope by placing an annular aperture in the front focal plane of the microscope condenser and a 'phase plate' at the back focal plane of the objective (Figure 10.16). The phase plate consists of a disc of glass on which is deposited a ring of some substance, such as magnesium fluoride, having a thickness such that light passing through it is retarded in phase by $\pi/2$ relative to light which passes through the glass alone. The system is arranged so that the illuminated image of the condenser annulus falls exactly on the ring

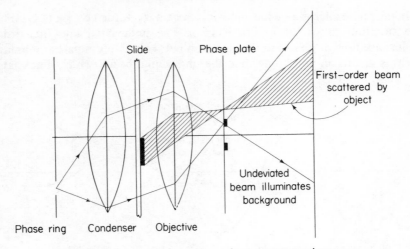

Figure 10.16.   Path of rays in a phase–contrast microscope

of the phase plate. Thus, in the absence of a specimen to scatter light, all light passing through the condenser annulus will pass through the ring of the phase plate. When a specimen is introduced light is scattered and will, for the most part, pass through regions of the phase plate other than the phase ring; undeviated light will pass *in toto* through the phase ring. The conditions of Figure 10.15(a) have thus been achieved since the undeviated beams will be retarded by $\pi/2$ with respect to the diffracted beams. If an annular absorbing film is superimposed on the area of the phase ring, the undeviated beams can also be reduced in intensity to correspond with the intensity of the diffracted beams.

Contrast in a phase microscope occurs only in regions where the optical properties of the specimen are changing rapidly, for example at the boundary

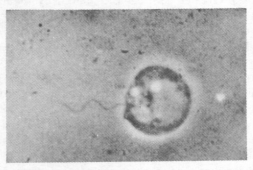

Figure 10.17.   Phase–contrast micrograph of a flagellated micro-organism. (Kindly supplied by Miss P.D. Peters, Queen Elizabeth College, London)

of a cell, and at such a region a halo is observed (Figure 10.17). This halo is an artifact and care must be taken over the interpretation of the image in the region of the halo. It can also be seen in Figure 10.17 that the brightness of the image in the centre of the specimen is roughly the same as that of the surroundings, even though the refractive indices of the two regions are different. Thus, the phase microscope should not be used for observing specimens where a gradual change in refractive index occurs.

10.7.3 *The Interference Microscope*

The images in the microscopes described so far are produced by interference between light beams scattered by the specimen. In the interference microscope the image is again formed by interference between light beams, but in this case the beams originate from a device below the specimen. The general principle is shown in Figure 10.18. Light from the condenser is

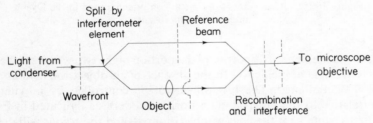

Figure 10.18.   Principle of the interference microscope

split into two beams, one of which passes through the object while the other is a reference beam and does not pass through the specimen. Beyond the specimen the two beams are recombined and interference occurs. The phase change introduced when light passes through the object produces a change of intensity in the image, thereby enabling a phase object to be detected. This system reveals gradual changes in the refractive index of a specimen, and the halo artifact characteristic of the phase–contrast microscope is not present. There are two basic types of interference microscopy, the double–beam method in which only two beams are used and the multiple–beam method where many beams are employed.

In the Dyson interference microscope (Figure 10.19) the light beam from the condenser is split into two by a glass interferometer plate with semi–silvered surfaces. A similar plate recombines the two beams and the composite beam is reflected at a spherical surface before entering the microscope objective. The interferometer plates are, in fact, wedges having sides inclined at a very small angle (about 5′ or 1·5 mrad). When the two wedges have the same orientation (Figure 10.20a), the path difference between the

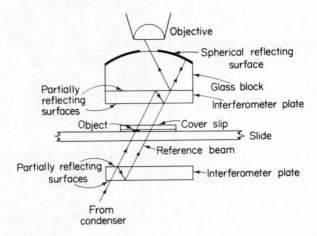

Figure 10.19.   Arrangement of optical components used in the Dyson
interference microscope

two beams is the same, irrespective of the portion of the two wedges traversed, for a given angle of incidence. In the absence of an object the field appears uniformly illuminated. The brightness can be controlled by moving one wedge relative to the other along a horizontal axis, as indicated in Figure 10.20(a). In white light the field will be coloured, and the colour will depend upon the relative positions of the two wedges. An object will appear in a

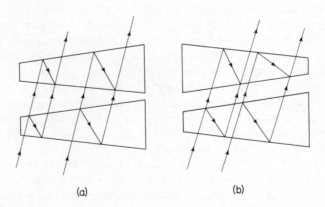

(a)   (b)

Figure 10.20.   Orientation of wedges to produce (a) a uniformly illuminated field and (b) fringes crossing the field of view. For clarity, the angles of the wedges in the diagram are much larger than the angles ($\simeq 5'$) found in actual wedges

contrasting colour which can be varied (at the same time as the background) by moving one of the wedges.

When the wedges are oppositely oriented, as shown in Figure 10.20(b), the path difference will vary according to the point at which the incident beam strikes the lower wedge. In this case the microscope field is crossed

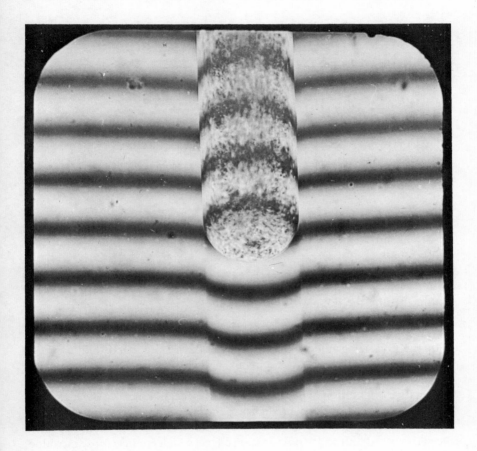

Figure 10.21. Interference micrograph of an amoeba pseudopod (which is distinguished by its granular appearance) moving in a narrow tube bored in agar. Water fills the other part of the tube. The optical path length for light which passes only through agar lies between the path lengths for light passing through the pseudopod or through water. Fringes in pseudopod and water are therefore displaced in opposite directions. (Kindly provided by Professor R.D. Allen, Department of Biological Sciences, State University of New York, Albany)

by fringes whose relative positions in object and background can be used to determine the thickness of the object if its refractive index is known (Figure 10.21).

In a multiple interference system the surface contours of a specimen can be examined with great precision. The method, one form of which is shown in Figure 10.22, can be used to detect depressions only one three–hundredth

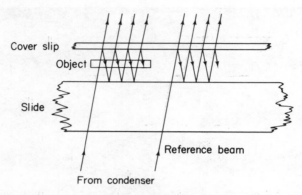

Figure 10.22.   Method for achieving multiple–beam interference

of the wavelength of light deep. However, the lateral resolution which can be achieved by this method is only about 5 μm compared with the 0·2 μm obtained using the double–beam technique.

10.7.4 The Polarizing Microscope

In the polarizing microscope the properties of polarized light are used to obtain information about specimens. A polarizer is usually placed below the microscope condenser while an analyser is placed above the objective. The relative orientation of polarizer and analyser can be accurately measured. Special objectives must be used in a polarizing microscope since any in-homogeneities in the lens (produced, for example, by strain occurring when the glass cooled) introduce polarization effects which confuse those introduc-ed by the object.

Perhaps the most important use of the polarizing microscope in biology is the detection and measurement of birefringence (Chapter 9). As noted in the earlier chapter, birefringence depends upon the arrangement of atoms and molecules within a substance, so that, for example, collagen, which has relatively unimportant side chains, has a different type of bire-fringence from a molecule such as DNA where the side chains are very important. A compensator, introduced between the polarizing devices of the microscope, can be used to describe the birefringence quantitatively in terms of the difference between the refractive indices for ordinary and extraordinary rays.

Further Reading

Casartelli, J.D. (1965). *Microscopy for Students*. McGraw–Hill, New York.
Françon, M. (1961). *Progress in Microscopy*. Pergamon, Oxford.
Kingslake, R. (Ed.) (1965). *Applied Optics and Optical Engineering*, Vol. 3. Academic Press, New York.
Martin, L.C. (1966). *The Theory of the Microscope*. Blackie, London.
Mollring, F.K. *Microscopy from the Very Beginning*. Carl Zeiss, Oberkochen.

PROBLEMS

10.1 Using the data provided in the chapter, estimate the minimum distance which can be resolved by a microscope when the object is illuminated by visible radiation.

10.2 The diatom *Nitzschia obtusa* bears a pattern consisting of many rows of fine dots. The rows are 2·5 μm apart, while the dots have a separation of 0·4 μm in each row. The specimen is illuminated by light of wavelength 546 nm and observed through a microscope. What is the minimum numerical aperture of the objective if (a) the rows and (b) the dots are to be resolved?

10.3 What is the minimum magnification which is required to render the dots of problem 2 visible? What would be the magnification of the eyepiece you would choose to observe the specimen if the objective had a magnification of × 50?

10.4 A particular 'dry' objective has a working distance of 1 mm. The diameter of the lens element nearest the specimen is 2·4 mm. If all the light from the specimen which strikes the front element contributes to the formation of an image, calculate the numerical aperture of the objective. What would be the effect on the numerical aperture of using this objective with an immersion oil of refractive index 1·5? (Assume that the working distance remains unaltered.)

What should be the numerical aperture of the condenser in the above instances if maximum resolution is to be derived from the microscope?

CHAPTER 11

The Electron Microscope

In the previous chapter we saw that in microscopes using visible light the resolution approaches about 200 nm (0·2 μm). Although this figure is adequate for the study of tissue morphology (the arrangement of cells) the light microscope is less useful for cytology (the study of a single cell and its internal structure) and of no use in trying to see objects as small as viruses. The scale of dimensions against which resolution can be measured is shown in Figure 11.1. Clearly, for cytology and virology and other studies at this structural level, an instrument is necessary which has a much better resolution than the light microscope. This need is fulfilled by the electron microscope, which has a practical resolution of the order of 0·5 nm (5 Å)

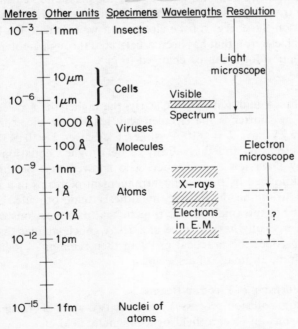

Figure 11.1.   A scale of dimensions for microscopy

293

and a theoretically obtainable one about a hundred times smaller still. The electron microscope, developed commercially since about 1940, has played a large part in the development of biology at a molecular level, both by allowing an insight into the detailed structures of the cell and by enabling biochemists to observe the effects of their methods.

11.1 Resolving Power

The best resolution obtainable with a microscope is given approximately by $\lambda/2$N.A., where λ is the wavelength of the radiation used and N.A. the numerical aperture of the system (Chapter 10). To get a resolution which will show large molecules it is clearly necessary to use wavelengths that are much shorter than those of light, and ideally of the order of magnitude of atomic dimensions. These wavelengths, and the resulting high resolution, are obtained in the electron microscope because it uses electrons instead of light. Although electrons appear in many cases to behave as particles, they also have wave–like properties (see the dual properties of a photon of light, Chapter 7), and in 1924 de Broglie suggested that the wavelength associated with an electron should be given by

$$\lambda = h/mv, \tag{11.1}$$

where h is Planck's constant and m and v the mass and velocity respectively of the electron. Equation (11.1) implies that λ depends on the momentum, and hence on the energy, of the electron; if we calculate the kinetic energy ($\frac{1}{2} mv^2$) of an electron that has been accelerated through a potential difference ϕ, the equation for λ can be changed to give

$$\lambda = \sqrt{150/\phi}, \tag{11.2}$$

where λ is in Å and ϕ in volts. Clearly, the higher the accelerating voltage becomes, the shorter is the electron wavelength (e.g. electrons accelerated through 15 kV have $\lambda = 0.10$ Å, while 75 kV gives $\lambda = 0.043$ Å). The wave nature of electrons and the validity of de Broglie's equation were proved in 1927 by Davisson and Germer, who showed that beams of electrons could be diffracted by the regular arrangement of atoms in a nickel crystal. It follows from the de Broglie hypothesis that, provided its numerical aperture is not too small, an instrument that uses electrons of high energy should theoretically have a value of $\lambda/2$N.A. much smaller than that of the light microscope; the practical problem then remains of focusing electron beams to form an image of a specimen.

11.2 The Focusing of Electron Beams

Apart from collision processes, there are two ways of changing the course of an electron: an electric field or a magnetic field can be used to deflect the electron path. Because of difficulties in engineering and electrical insu–

lation, most modern electron microscopes use magnetic fields for focusing purposes, rather than electrostatic ones. However, all instruments use an electrostatic field to produce a beam of electrons, and the electrons are also subject to electrostatic forces at various points on their path through the microscope.

11.2.1 *Electrostatic Fields*

Let us consider an electron of charge $-e$ which starts at rest from a point where the electrostatic potential (Chapter 4) is zero. If the electron moves to a point where the potential is ϕ, the sum of its kinetic and potential energies remains the same as the initial value (i.e. 0) since energy is conserved, and we can write

$$\tfrac{1}{2}\,mv^2 + (-e\,\phi) = 0. \tag{11.3}$$

(In conventional electron microscopes where the voltages are less than 100 kV we can neglect the fact that m depends on v because of relativity and can assume that m is constant for the electron.) The momentum mv of the electron is thus given by

$$mv = \sqrt{2me\phi} \tag{11.4}$$

after manipulating equation (11.3). Insertion of the momentum into the de Broglie equation (11.1), together with known values of m and e for the electron, leads to equation (11.2) for the electron wavelength.

We are now in a position to see what happens to an electron as it passes through an electric field. (Qualitatively, the electrons in such a field tend to move at right angles to the equipotential lines, in the way that a stream tends to run downhill at right angles to the contour lines of a relief map.)

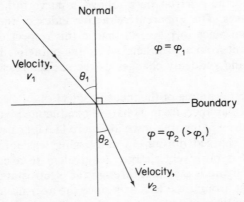

Figure 11.2.   Electron path 'refracted' at a boundary between regions of different electrostatic potential

Suppose, as in Figure 11.2, an electron that started at rest where ϕ is 0 is passing through a region where there is a uniform potential ϕ_1 and then crosses a boundary where the potential changes suddenly to ϕ_2. The path of the electron appears to be 'refracted' towards the normal to the boundary and the deflection of the path can be found by using the fact that in a direction parallel to the boundary the momentum of the electron is conserved, since there is no force on the electron in that direction. Taking components of the electron momentum parallel to the boundary we have

$$mv_1 \sin \theta_1 = mv_2 \sin \theta_2,$$

that is

$$\frac{\sin \theta_1}{\sin \theta_2} = \frac{mv_2}{mv_1}$$

$$= \frac{\sqrt{2me\phi_2}}{\sqrt{2me\phi_1}}, \text{ from equation (11.4).}$$

Thus, the apparent 'refractive index', n, of the boundary is given by

$$n = \frac{\sin \theta_1}{\sin \theta_2} = \sqrt{\frac{\phi_2}{\phi_1}} . \tag{11.5}$$

In other words, the electron path is bent as it travels through regions where the potential is changing, and the changes in apparent 'refractive index' are greatest for the region where the potential varies most rapidly with distance. Note that the path is refracted towards the normal when going to a region of higher potential.

In reality, there are no regions in an electrostatic field where the potential jumps from one constant value to another in the way shown in Figure 11.2. This figure was an approximation to the true situation, in which ϕ varies smoothly and the electron path follows a curve rather than a succession of straight lines. The apparent refractive index of the field in fact varies smoothly from place to place, so that in this respect the electrostatic field is very different from a glass lens, where the refractive index of the medium is constant, and suddenly changes as the surface is crossed between glass and air.

The simplest example of the focusing effect of an electrostatic field is the effect of a circular aperture in a charged conducting plate. The equipotential lines and electron paths are shown in Figure 11.3 for cases where the aperture is negatively charged (forming a converging lens for the electron beam) and where it is positively charged (equivalent to a concave lens in light optics). Each figure is a *section* through the electrostatic field, which in fact is symmetrical around the lens axis since the aperture is circular. It can be proved, in fact, that any electrostatic field that is axially symmetric acts as an ideal lens for paraxial electron paths.

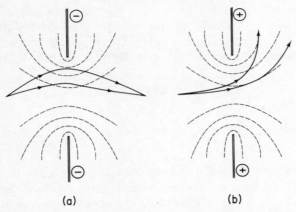

Figure 11.3. Charged circular apertures as electrostatic lenses (shown in cross-section): ------ indicates equipotential lines and →— electron trajectories. (a) A negative converging aperture and (b) a positive diverging aperture

11.2.2 *Magnetic Fields*

When a moving electron has a velocity perpendicular to the direction of a magnetic field, the field exerts a force **F** on the electron in the third perpendicular direction, as shown in Figure 11.4. The force produces a

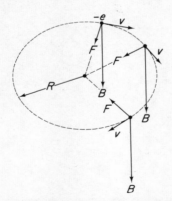

Figure 11.4. An electron in a constant magnetic field **B** is constrained to move in a circle by the centripetal force, **F**, if v is normal to **B**

sideways acceleration of the electron which constrains the electron to follow a circular path, if no other forces are present. For the situation shown,

$$F = Bev \text{ (newton)}, \tag{11.6}$$

where B is the field in tesla (Wb m^{-2}), e the electronic charge in C and v the velocity in m s^{-1}. Obviously, the stronger the field, the stronger the centripetal force F becomes and the tighter the curve of the circling electron. From the principles of circular motion (Chapter 1) we can deduce that

$$\frac{mv^2}{R} = Bev, \tag{11.7}$$

that is

$$R = \frac{mv}{eB}. \tag{11.8}$$

Now if an electron is moving at some skew angle (i.e. not 0 or 90°) to the magnetic field lines, we can resolve its velocity into one component parallel to the field and one at right angles to it. The parallel component describes the motion of the electron along the field lines, while the other component leads to the circular motion described above. The combination of the two motions produces a spiral path, and, if the field is uniform (Figure 11.5)

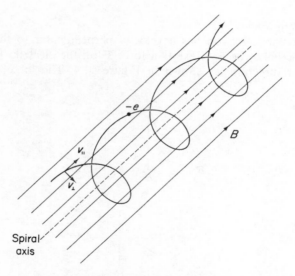

Figure 11.5.   Path of an electron in a uniform magnetic field

and the velocity of the electron is constant, the radius of the spiral remains constant so that the electron never crosses the axis of the spiral (i.e. it never comes to a focus). Thus, in general, to focus a beam of electrons and make the ray paths cross through one point, a *non*-uniform field is necessary.

The non–uniform field which acts as a magnetic lens for electrons is usually produced by what is essentially an electromagnet with specially shaped north and south poles. Any lens system must be symmetrical about the lens axis, so the poles are made in the form of rings or hollow cones of iron alloy as shown in Figure 11.6. To obtain the short focal lengths which

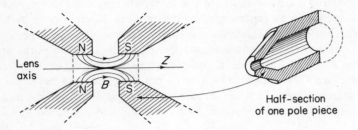

Figure 11.6.   Poles of a magnetic lens showing the non–uniform field

are needed for high magnification, the gap between the N and S pole pieces may be only a few millimetres and large magnetic fields are produced of the order of 1 tesla (10^4 gauss or $1\,\mathrm{Wb\,m^{-2}}$).

The focusing action of the magnetic field in the gap between the pole pieces is extremely complicated. For rays which pass through the field very close to the axis and at small angles, the field acts as an ideal lens for electrons and a focal length can be computed if the magnetic field B_z along the axis is known at every point z on the axis of the system. The variation of B_z through the lens is shown in Figure 11.7; the focal length of a 'thin' lens,

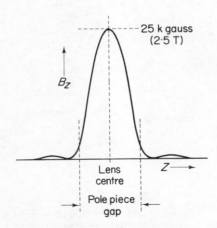

Figure 11.7.   Variation of the axial magnetic field B_z along the axis of a magnetic lens. (The field is not uniform, and rises to a maximum value.)

where the pole–piece gap is much smaller than the focal length, is given by

$$\frac{1}{f} = \frac{e}{8m\phi} \int_0^l B_z{}^2 \, dz, \qquad (11.9)$$

where 0 and l are the limits of the lens. Whereas the focusing action of an optical lens can be represented by a single ray diagram on a plane piece of paper, this is not possible for the magnetic lens because, superimposed on the focusing action, there is a rotation of all the electron paths around the axis of the lens, rather as if the plane piece of paper were twisted in the middle (see Figure 11.8). This occurs not only for skew rays, as one might expect from Figure 11.5, but for all rays which pass through the non–uniform field, so that the image formed by a magnetic lens is rotated, relative to the

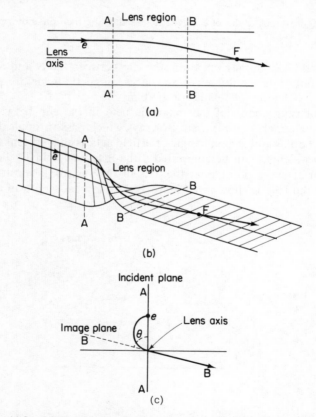

Figure 11.8. (a) Focusing action of a magnetic lens for a parallel ray (represented on a plane strip). (b) Focusing action of a magnetic lens in three dimensions. (c) Motion of the electron in (b) viewed along the axis. The image plane is rotated through an angle θ

object, not by 180° as in a glass lens, but by an angle which varies with the strength, and thus the focal length, of the lens.

One difference between the magnetic lens and the glass one is that in the former its focal length can be changed continuously, at will, by changing the magnetic field. In practice, a real, 'thick' magnetic lens (to which equation (11.9) does not apply) has a minimum focal length which is about half the gap between the pole pieces; at this maximum strength of the lens, the image rotation is about 115° (2 rad).

If we examine the simple equation (11.9), various general properties of the magnetic lens become evident. First, we see that the focal length f depends on $1/B^2$. This means that a strong field produces a short focal length and also implies, since the field appears as a squared term, that f must always be positive. A magnetic lens, therefore, is always a converging lens, like a convex optical lens. We can also see that f depends, through ϕ, on the energy of the electrons, but since λ is also related to ϕ (equation 11.2) this means that f depends on λ. The focal length of the lens varies with the wavelength, in other words, and since this defect is called chromatic aberration in glass lenses the term is used to describe this defect of the magnetic lens, too, although there is nothing chromatic about electrons. Finally, we see that the focal length depends twice as much on B as on ϕ in the equation, since B is squared and ϕ appears only to the first power. This means that to produce a magnetic lens with the steady value of f which is necessary to form a clear image, the electric currents which produce the magnetic field must be twice as free from fluctuation as the accelerating voltage of the instrument.

11.3 Lens Construction and Aberrations

11.3.1 The Magnetic Lens

As mentioned previously, the magnetic lens is essentially the region between two conical pole pieces which are spaced a few millimetres apart on the axis of the instrument. An electromagnet has to be designed to concentrate a large magnetic field between the poles, and this arrangement, shown in Figure 11.9, is also known in toto as a magnetic lens. The pole pieces, of magnetic alloy, are separated by a spacing collar which is non-magnetic, so that the magnetic flux produced by the current in the lens windings is concentrated into the pole-piece gap and does not 'short-circuit' it by going through the spacer. The flux is also concentrated by surrounding the lens coil, inside and out, with a 'soft' iron casing consisting of inner and outer cylinders and circular plates at top and bottom. In some lenses the pole pieces are removable so that ones of different design can be inserted for high or low ranges of magnification. Weak lenses of long focal length may not need separate pole pieces to concentrate the magnetic field, and

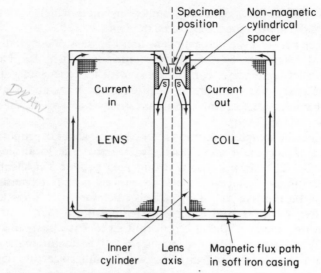

Figure 11.9.   Objective lens (schematic) of an electron microscope
(longitudinal section)

in this case the flux path through the inner cylinder of the lens casing is interrupted only by a non–magnetic spacer to form the 'gap' (see Figure 11.10). Obviously, the controlled movement of such heavy lens systems would be difficult, so it is fortunate that the lenses can be focused by altering

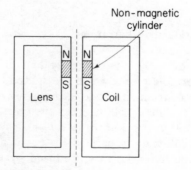

Figure 11.10.   A lens of long focal length (e.g. condenser lens), without
pole pieces

their focal lengths rather than by moving the lenses bodily. As we have seen, the lens current determines the magnetic field and focal length of the lens, and to obtain a focal length constant enough to give a resolution of 10 Å in the image, the current must be stabilized so that any A.C. ripple amounts to less than ten parts per million of the steady current.

11.3.2 Lens Aberrations

The magnetic lens suffers from the same geometrical aberrations that occur in glass optics, but because the aperture of the lens must be small (as we shall see) the off–axis aberrations such as coma, astigmatism and curvature of field (see Chapter 8) are negligible, and distortion (of the pin–cushion kind) only appears at low magnifications, becoming minimal at the high magnifications for which the lens is designed. The main defect of the lens is, in fact, the remaining one of the five, that is *spherical aberration*. Since all magnetic lenses are convergent the spherical aberration is always positive, that is the lens power becomes greater in the outer zones of the lens and rays travelling through these zones come to a shorter focus than rays near the axis. A point object gives a disc of least confusion instead of a point image, and the radius of the disc, which effectively limits the resolution of the image, corresponds to a distance in the object of Δr, where

$$\Delta r = C_s \alpha^3. \qquad (11.10)$$

In this equation α is the semi–angular aperture of the lens (see Figure 11.11)

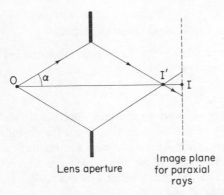

Figure 11.11.   Spherical aberration. I′ is the image of O formed by rays at an angle α to the axis; I is the image formed by the centre of the lens

and C_s a spherical aberration constant for the lens which has a value in the region $0.3 \, f$ to $1.0 \, f$. To improve the resolution in a given lens it is clear that α must be reduced. This is particularly important in the objective lens, which provides the first stage of magnification in the instrument; this lens is provided with an *objective aperture* (usually a molybdenum disc with a pinhole at its centre), which limits the angle of electron paths entering the lens from the specimen. The aperture is either positioned in the pole–piece gap or behind the lens at the back focal plane (Figure 11.12).

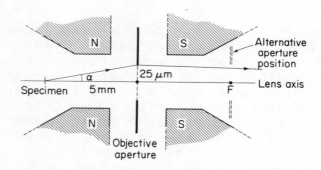

Figure 11.12. Schematic diagram of objective lens and aperture (typical values)

The existence of the aperture causes other resolution problems, because in effect the aperture cuts down further the information passed by the lens to the image, so that the resolution may then become limited by diffraction at the aperture. The resolution limit in the object, because of diffraction, was given in Chapter 10 and is

$$D_{min} \simeq \lambda/(2 \sin \alpha),$$

since $\sin \alpha$ corresponds to the numerical aperture of the magnetic lens. We can simplify further by writing

$$D_{min} \simeq \lambda/(2 \alpha) \tag{11.11}$$

for the small angles allowed by the lens aperture. If we compare equations (11.10) and (11.11), it can be seen that if we try to improve the resolution by decreasing α and thus Δr we increase the resolution limit, D_{min}. A compromise is called for, at a value of α which reduces Δr without increasing D_{min} excessively. The optimum value for α is found to be one where D_{min} is about equal to Δr, that is

$$C_s \alpha^3 = \lambda/2\alpha. \tag{11.12}$$

In practice, the solution of equation (11.12) leads to values of α of about 5×10^{-3} rad and objective aperture diameters of around 50 μm.

Apart from the geometrical aberrations common to spherical glass lenses the magnetic lens has defects which are inherent in its nature: the more important ones are chromatic aberration and axial astigmatism. In addition, the resolution is impaired by instabilities in the lens current and the accelerating voltage of the microscope. The *chromatic aberration* of the lens has been referred to already, in the discussion of equation (11.9) for the focal length. Essentially, the focal length depends on the electron energy and because of this anything that affects the latter in a non–uniform way also impairs the resolution. The resolution limit is given by

$$\Delta r = C_c \alpha \left(\frac{\Delta \phi}{\phi} \right), \tag{11.13}$$

where C_c is a constant for the lens and $(\Delta \phi / \phi)$ is the fractional fluctuation in electron energy. Energy variations are introduced by ripple in the high–voltage supply which accelerates the electrons to their final velocity, and, to a small extent, by the thermal energies acquired by the electrons as they leave the hot tungsten filament which acts as the electron source. Another more troublesome variation is due to energy losses suffered by electrons as they collide with atoms of the specimen—a process known as inelastic electron scattering. The losses result in a longer wavelength of the electron, which is then no longer focused to the correct image point because of the chromatic aberration of the lens. A thick specimen in which many losses occur gives rise to a background 'fog' of badly focused electrons which reduces the image contrast and also the resolution. The remedy is to use thin specimens or sections of specimens, and it has been calculated that the specimen thickness should be less than twenty times the resolution required. This implies that to resolve 10 Å a section thickness of less than 200 Å is required; special machines (ultramicrotomes) are needed to cut such sections, which are thinner than a tenth of a wavelength of visible light.

The final aberration that we should consider, which is of much practical importance, is *axial astigmatism*. This is due to imperfections in the magnetic pole pieces, caused either by magnetic inhomogeneity of the metal or by defects in machining the shape to the close tolerances required. The imperfections result in an asymmetrical lens field which gives the effect of a weak cylindrical lens superimposed on the spherical–lens effect of the perfect lens. This, in turn, means that the magnetic lens has focal lengths of different

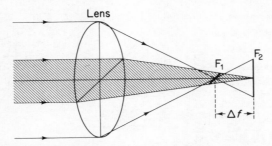

Figure 11.13.   Axial astigmatism gives two line images of a point object. Δf is the 'longitudinal astigmatism'

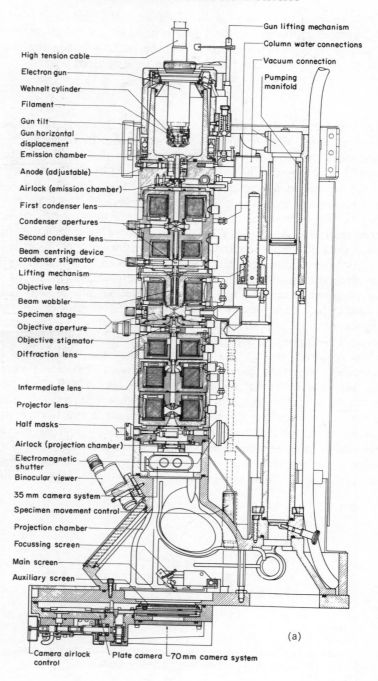

High tension cable

Electron gun

Wehnelt cylinder

Filament

Gun tilt

Gun horizontal displacement

Emission chamber

Anode (adjustable)

Airlock (emission chamber)

First condenser lens

Condenser apertures

Second condenser lens

Beam centring device condenser stigmator

Lifting mechanism

Objective lens

Beam wobbler

Specimen stage

Objective aperture

Objective stigmator

Diffraction lens

Intermediate lens

Projector lens

Half masks

Airlock (projection chamber)

Electromagnetic shutter

Binocular viewer

35 mm camera system

Specimen movement control

Projection chamber

Focussing screen

Main screen

Auxiliary screen

Gun lifting mechanism

Column water connections

Vacuum connection

Pumping manifold

Camera airlock control

Plate camera

70 mm camera system

(a)

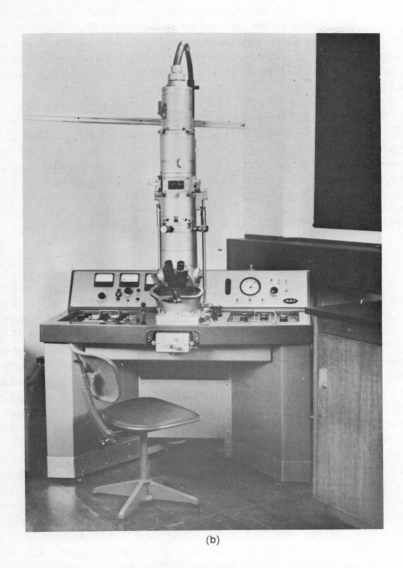

(b)

Figure 11.14. (a) Cross–section of a Philips EM 300 electron micro–scope. (Kindly supplied by Pye Unicam Ltd., Philips Analytical Department, Cambridge) (b) A photograph of an A.E.I. EM 6 electron microscope. (Reproduced with kind permission of A.E.I. Scientific Apparatus Ltd., Harlow, Essex)

values for different axial planes: the focal length is a maximum for a particular plane and a minimum for the plane at right angles to it (Figure 11.13). To correct this effect, modern objective lenses are fitted with electrostatic devices (called, unfortunately, 'stigmators') which produce a cylindrical–lens effect in opposition to the one inherent in the objective. The strength and orientation of this compensating lens effect can be controlled by the operator of the electron microscope so that it exactly removes the astigmatism of the image.

11.4 Construction and Operation of the Electron Microscope

The electron microscope is a complex instrument and requires in its design the services of the vacuum engineer, the electronics expert, the electrical engineer and a designer of fine mechanical movements. Most electron microscopes are built with the lenses mounted vertically above each other in a column, rather than with the lens axis horizontal. A typical example is shown in Figure 11.14(b) and a labelled cross–section is given in Figure 11.14(a). The functions of the various parts of the microscope are discussed below.

11.4.1 *Vacuum System*

The interior of the electron microscope is evacuated to a low pressure of about $0.5\ \mu m$ of mercury ($0.67\ N\ m^{-2}$) when in use and is left under vacuum at other times to reduce the adsorption of air and water molecules on the inner surfaces. The vacuum is necessary for four reasons. First, it serves to insulate the high voltage of the electron source from the rest of the microscope. Secondly, the vacuum reduces the chance that the electron paths will be disorganized by collisions with gas molecules, thus impairing resolution. Thirdly, because the electron source is a white–hot tungsten filament, a vacuum is necessary to preserve the tungsten from oxidation— the length of life of the filament is approximately inversely proportional to the operating pressure in the microscope. Lastly, since the presence of water and various other molecules results in contamination of the specimen, it is necessary to reduce vapour pressures in the instrument to a minimum. Specimen contamination becomes a serious problem when taking high-resolution pictures of particles in the electron microscope. Under bombardment by the electron beam, hydrocarbon molecules in the atmospheric environment of the specimen form polymers on the specimen surface, so that a layer of contamination can grow at the rate of from $1\ \overset{\circ}{A}$ to $10\ \overset{\circ}{A}$ per second. In addition, ionized water molecules can have an etching effect on organic specimens.

The vacuum system of the electron microscope is designed so that the vertical column of the instrument (Figure 11.14a) can be evacuated 'roughly' (i.e. to a fairly low pressure) with a rotary pump connected by a manifold system to various parts of the column. When the column is under vacuum the pressure is further reduced by an oil diffusion pump; the rotary pump is now used to remove the exhaust from the diffusion pump, instead of pumping through the manifold. A moisture trap, consisting of a tray of phosphorus pentoxide, is often placed in the vacuum system to absorb water vapour, and care is taken to use in the system only oils and greases of very low vapour pressure so as to reduce the specimen contamination rate. Most electron microscopes have air–locks for introducing a specimen or a camera cassette without admitting air to the whole of the column. This saves time in regaining the operating vacuum and ensures that most of the microscope is always at reduced pressure.

11.4.2 *Electronic System*

The purposes of the electronic system of the microscope are (a) to provide a high voltage of up to 100 kV, stabilized within about twenty parts per million, with a choice of two or three operating values; (b) to provide lens currents of the order of 100 mA, stabilized to five parts per million and continuously variable so that the lenses can be focused; and (c) to provide a heating current for the filament which is the electron source. In addition to the electronics there is, of course, an electricity supply to operate the vacuum system, signal lights and safety relays, etc. The various lens currents and the high–voltage supply are partly stabilized (or 'smoothed') by conventional smoothing circuits, and the outputs from these circuits are further regulated to an extremely steady value by the use of *negative feedback* (see Chapter 6). In essence, a fraction of the voltage to be regulated (or a voltage proportional to the current to be regulated) is compared electronically with an extremely steady *reference voltage* which is usually provided by dry batteries. Any difference between the two values is detected and used to control the regulated value so as to reduce the difference to zero. A schematic diagram of a system for regulating lens currents is shown in

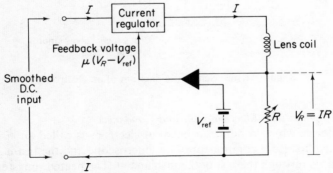

Figure 11.15.   Lens–current regulation by negative feedback

Figure 11.15. The voltage across R, which is proportional to the lens current, I, is compared with the battery voltage, V_{ref}, and an amplified voltage proportional to $(V_R - V_{ref})$ is used to control the current I through the regulator circuit. If $(V_R - V_{ref}) > 0$ the regulator reduces the value of I, and if $(V_R - V_{ref}) < 0$ the value of I is increased—in either case this brings $(V_R - V_{ref})$ back to zero and hence maintains a steady value of I. The value of the stabilized current can be increased by reducing R, since the feedback circuit then allows a larger current through the smaller resistance to maintain the same voltage (IR) across R as before. The resistance R is thus the focusing control for the lens.

The high–voltage supply for the microscope is produced by taking the current from a high–frequency oscillator and passing it through a step–up transformer (Chapter 5) to provide an alternating high voltage. This is rectified and further increased by voltage–doubling circuits comprising capacitors and diodes, and is then smoothed and regulated. The high–voltage supply (negative with respect to earth) and the current for the filament, which is at nearly the same potential as the cathode cap (Figure 11.16), are carried to the top of the electron microscope column through a highly insulated cable.

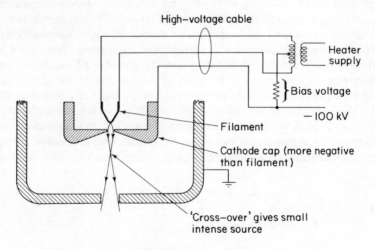

Figure 11.16.   Schematic diagram of an electron gun

11.4.3 Illumination: Electron Gun and Condenser

The device which produces a beam of electrons is called an *electron gun*. This is the first (top) component of the microscope and the beam is directed downwards through the rest of the instrument. The heated tungsten filament in the gun is maintained at a high negative voltage so that electrons ejected from the filament 'see' the rest of the instrument (at earth potential) as being highly positive with respect to their origin. The electrons are thus accelerated downwards into the instrument by this potential difference. In order to collect as many electrons as possible from the filament and produce an intense, small, electron source, a 'cathode cap' is positioned in front of the filament and is maintained more negative than it by a few hundred volts (Figure 11.16). A circular aperture in this negatively biased cap acts as a converging electrostatic lens (Section 11.2.1) so that the electrons are focused to a 'cross–over'. The electrons from this effective source then accelerate through an aperture at earth potential (the *anode*, since it is 100 kV more positive than the filament), which acts as a slightly diverging lens. The electrons thus pass through the anode at their maximum velocity in a slowly diverging beam.

The beam from the electron gun next enters a magnetic *condenser lens* which serves to control the 'illumination' of the object by concentrating the electron beam with more or less intensity upon the specimen. In effect the condenser lens can produce a focused image of the gun cross–over upon the specimen and in this situation the beam current density ($A\ m^{-2}$) is most intense, since all the electrons pass through a small area of the specimen. If the condenser lens is defocused by altering the lens current, the image of the cross–over moves above or below the specimen, which is then irradiated over a larger area at a lower intensity. Often double condenser lenses are used, which give greater flexibility in controlling the illumination and enable the area irradiated at the specimen to be reduced to a spot of from 2 μm to 3 μm diameter instead of the 20 μm to 50 μm obtained with a single lens. This gives an increased maximum intensity and confines the effects of specimen contamination to a smaller area.

As in the optical microscope, the conditions of illumination are of critical importance in achieving the best resolution of the instrument. The optimum, theoretical resolution of which the electron microscope is capable is calculated on the assumption that the specimen is illuminated with a coherent beam of electrons. In practice, this coherence is partially obtained by ensuring that the electrons come from a region which approximates as nearly as possible to a point source and by keeping the convergence of the illuminating beam as small as possible.

11.4.4 *Specimen Chamber and Subsequent Lenses*

Below the condenser lens lies the specimen chamber, with a door or air–lock through which the specimen is introduced to lie in its position just above the upper focal point of the objective lens. The specimens are mounted on thin, circular perforated discs called 'grids', usually of copper and about 3 mm in diameter, and the specimen stage in the chamber must be capable of traversing a specimen grid across the field of the objective lens with the very fine movement demanded by high powers of magnification. Two types of specimen holder are shown in Figure 11.17: in (a) the grid is carried on the end of a tube inserted from above the lens, while in (b) several grids are carried over holes in a rod that passes through the pole–piece gap,

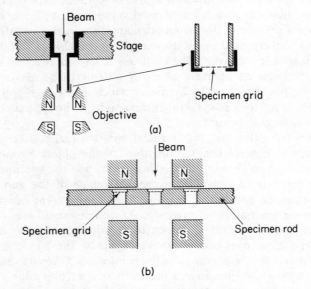

Figure 11.17.   Specimen holders

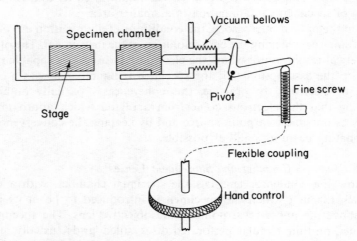

Figure 11.18.   One type of specimen–stage control (only one movement is shown)

thus allowing a number of specimens to be examined one–by–one. In (a) the stage is a copper block which slides on the floor of the specimen chamber and is traversed by push–rods which enter the vacuum system through metal bellows (Figure 11.18), while in (b) a more complicated arrangement is necessary to give the specimen holder the required movements.

Special specimen stages have been made with great ingenuity for examining heated or cooled specimens or specimens subjected to tension, etc., and one which deserves description here is a stage for stereoscopic electron micro–graphy—the production of three–dimensional images by electron micro–scopy. For stereoscopic pictures to be taken a tilting specimen stage is necessary which allows the normal to the specimen grid to be tilted through about 10° (0·17 rad) on either side of the lens axis. Photographs of the specimen ('electron micrographs') are taken for both tilted positions, and the great depth of field of the objective lens allows a reasonable area of the specimen to remain in focus despite the tilting. The two electron micrographs are then mounted as a stereo pair and projected or viewed in the usual way to give an apparently three–dimensional image of the specimen.

The objective lens, which has already been described in some detail, is immediately beneath the specimen chamber and serves to form a magnified intermediate image of the specimen in exactly the same way as the objective of a light microscope. The specimen lies just above the first focal point and close to it, so that a highly magnified image results. The lens has a removable objective aperture which fulfils the double purpose of limiting the aberrations of the lens and increasing the image contrast, as we shall see later. Since the aperture may be about 25 μm in radius and the focal length of the lens around 5 mm, the semi–angular aperture, α, is of the order of 5×10^{-3} rad, or 0·3°.

The intermediate image formed beneath the objective is further magnified by two projector lenses which project the final image on a fluorescent viewing screen. The projector lenses can be seen to be equivalent to the eyepiece lenses of an optical microscope. In the light microscope the magni–fication can be changed by inserting a different objective lens, although with a single objective some alteration can be obtained by using different eyepieces. In the electron microscope the focal length of the objective is usually fixed and magnification can therefore only be controlled by the projectors. To obtain low magnifications the two projector lenses are used in 'opposition' with one acting as a demagnifying lens, while to obtain high values the lens currents (and focal lengths) are arranged so that the lenses magnify successively. The range of magnifications that can be obtained with a given electron microscope may be as much as from 1 000 to 160 000 times, although in some instruments a change of objective pole pieces may be necessary at some point in the range, in addition to the variation of lens currents.

Another function of the projector lenses is to turn the microscope into an *electron–diffraction* camera if so required. If the focal lengths of the projectors are appropriately altered, a pattern can be obtained on the fluorescent screen which is an image of the back focal plane of the objective lens, rather than of the intermediate image plane. Since a diffraction pattern of the specimen is formed in the focal plane of the objective, a magnified image of this pattern is seen on the fluorescent screen. In this way the electron microscope can be used to study the structure of crystalline materials by electron diffraction in a manner analogous to the technique of X–ray diffraction.

11.4.5 *Image Recording*

In the conventional electron microscope the electron image formed by the projector lenses is made visible by a fluorescent screen. This is a metal plate coated with a layer of phosphors, such as zinc sulphide, which are excited by the bombarding electrons and re–emit the excitation energy (Chapter 7) as photons of visible light. A yellow–green phosphor is used because at this colour the eye has its peak sensitivity. The resolution of the screen is limited by the size of the phosphor particles and is about 100 μm. The screen is viewed from outside the viewing chamber through binoculars which allow the eye to resolve the grain of the phosphor; the eye is then receiving the maximum information obtainable from the image recorded by the screen. After focusing the electron image by this means the fluorescent screen can be moved away to allow the electrons to fall on a photographic plate or film which lies in a cassette below. The image which is recorded in this way is called an *electron micrograph*. Since the resolution of a photographic emulsion is about 1 μm, it records much more detail than the fluorescent screen. This is the reason for focusing the image on the screen as carefully as possible—a departure from focus which is only just apparent on the screen becomes extremely obvious in the electron micrograph. You may wonder why an image focused on a screen should still be in focus on a photographic plate some way (possibly 100 mm) below it; the answer lies in the extremely small aperture (about $0.3°$) of the electron microscope system, which leads to a very great depth of focus in the final image.

The advantage of the electron micrograph is that it is a permanent record of high resolution, and negatives showing a direct magnification of up to 100 000, say, can still be enlarged photographically to yield final magnifications of about 10^6. The disadvantage of the conventional method of image recording is in the visual system of focusing with a fluorescent screen, which becomes very difficult at high magnifications where the image intensity is consequently low. In order to improve on this system some electron microscopes have been fitted with an electronic *image intensifier* instead of a screen and camera; the apparatus is similar in principle to a television camera although it converts electrons, rather than photons, into a scanned television image. The brightness can be amplified electronically, so that the associated television screen can be viewed at a comfortable light level and used instead of the conventional fluorescent screen. The disadvantage of this method is the present lack of resolution in the scanned image, when compared with a photographic emulsion.

11.5 Image Contrast

11.5.1 Physical Basis of Contrast

One of the important differences between the light and electron microscope is in the way an object has its effect on the brightness or intensity of the corresponding image. Although in both microscopes we can achieve inter-ference effects (since both photons and electrons have wave properties), these effects are chiefly seen when the image is not in focus and consist of apparent fringes round the edges of opaque objects. However, the main process by which an object is recognized in transmission microscopy is by its effect in reducing the intensity of the radiation which reaches the corres-ponding part of the image. This process is different in the two kinds of microscope.

In the case of the ordinary light microscope, photons which pass into the specimen are either transmitted, scattered or totally absorbed and converted into other forms of energy. Because of the very large numerical aperture of the objective in this microscope, most of the scattered light is collected by the lens and focused to the correct points of the image; the only light which does not reach its destination is the component which has been absorbed by the specimen. Thus, in this case, it is *absorption* which diminishes the image intensity (i.e. a strongly absorbing object gives a dark image) and scattering is of little importance, except in special instru-ments like the phase–contrast microscope.

The electron microscope, on the other hand, uses electrons which, unlike photons, are very rarely absorbed in the sense that they cease to exist as particles. Thus the main process by which the intensity of an electron beam is diminished is by the scattering of electrons out of the beam after their interaction with atoms in the specimen (the process of *electron–scattering*). Now the electron microscope objective has a semi–angular aperture of about 0·3° while a light microscope objective may have a corres-ponding value of 60°, so that although in the latter case scattered light reaches the image, the scattered electrons do not, for the most part, reach the image in the electron microscope. They are prevented from contributing to the image intensity because the objective aperture intercepts all except those scattered at a minimal angle to the beam (Figure 11.19). A dark image on the fluorescent screen is therefore due to part of the specimen which

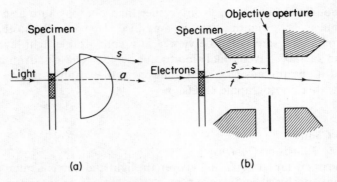

Figure 11.19. Comparison of contrast production in the light (a) and
electron (b) microscopes. In (a) s is scattered light and a is light attenuated by absorption, while in (b) s stands for scattered electrons lost from
the image and t for transmitted electrons

scatters electrons strongly. Obviously, a small aperture which intercepts
most of the scattered electrons will make a scattering region of the specimen
appear darker than if a large aperture were used which allowed more
scattered electrons to reach the image; the objective aperture size thus
affects the *contrast* of the final image.

11.5.2 *The Effect of the Specimen*

If I is the intensity of a given region of the electron image and I_0 the
intensity corresponding to, say, a hole in the specimen where no electrons
have been scattered, then the contrast of the given region can be defined
as $\ln(I_0/I)$. It can be shown that the contrast is dependent on the mass
per unit area W which the specimen presents to the electron beam, and that
for a region consisting of a single element,

$$\ln(I_0/I) = S_t W. \tag{11.14}$$

Here W is in kg m^{-2} and S_t is a measure of the efficiency of the material
in scattering electrons at angles greater than the objective aperture. The
value of S_t theoretically depends on the size and nature of the particular
atom, the electron velocity and the aperture angle. Electron–scattering can
be due to either elastic or inelastic collisions of the electrons with atoms,
and while inelastic events are important for light atoms elastic collisions
predominate with heavy ones. The combined effect of both kinds of collision
is included in the effective value of S_t, which in practice is sensibly constant
over the range of elements and depends more upon the characteristics of
the electron microscope (accelerating voltage, objective aperture, etc.) than
on the composition of the specimen. This implies that contrast in the electron

microscope image depends, as far as the specimen is concerned, only on its mass per unit area or *mass–thickness*, W. A typical value of S_t for an instrument working at 50 kV may be around 10^4 m^2 kg^{-1}.

The value of S_t can be used to deduce the maximum thickness of specimen that is useful in the electron microscope. If the specimen is too thick, the intensity of the image is reduced to a point where detail is no longer visible. This occurs when $\ln(I_0/I)$ is about 2·3 and I is a tenth of the value, I_0, obtained without a specimen. Then, taking the value of S_t already quoted, we have from equation (11.14) that $2·3 = 10^4\ W$ and W is $2·3 \times 10^{-4}$ kg m^{-2}. This is the maximum mass–thickness according to our assumptions, and for a substance with the same density as water corresponds to a thickness of $2·3 \times 10^{-7}$ m or 230 nm. (Although this is the maximum thickness at which some detail might be seen, of course the thickness must be about ten times less to obtain useful resolution, because of chromatic aberration.)

11.5.3 *Methods of Altering the Contrast*

It has already been explained that S_t, which governs the contrast in a given instrument, depends on the objective aperture value, so it might be thought that varying this would give enough control to make a thick specimen appear more transparent, or to give a thin one better contrast. However, the aperture can only be varied slightly without affecting the resolution of the instrument, so that other methods have been sought for changing both S_t and W. One other way to change S_t is to change the electron velocity, and both low– and high–voltage electron microscopes have been built for this purpose. The former gives a large value of S_t and is theoretically suitable for examining very thin specimens which need extra contrast, although practical difficulties have limited its development. The high–voltage electron microscope, on the other hand, gives a small value of S_t and is eminently suitable for very thick specimens which would otherwise appear opaque. Machines have been built which work at 1 MV or more, instead of the conventional 100 kV, and have the advantage that the reduced electron–scattering in this case results in less damage to the specimen. The greater penetration which is possible with these machines makes it likely that whole living cells might be examined, enclosed in a liquid environment, instead of the dehydrated thin sections observed at the present day.

Finally, although general specimen techniques are not within the scope of this book, one must mention the third way of changing contrast which, in scientific jargon, is called *'electron–staining'*. In light microscopy a wide variety of selective staining procedures is available for giving extra contrast, in colour, to specific substances in the specimen. The comparable activity in electron microscopy is to try to increase the value of W of selected regions by attaching heavy atoms to a particular substance. There is no colour, of course, and one can only make regions appear blacker with the so–called 'electron stains', which consist of chemicals such as mercuric chloride, phosphotungstic acid and uranyl acetate, containing respectively the heavy atoms mercury, tungsten and uranium. Electron–staining to show specific substances is still in its infancy, but some advances have been made in detecting the sites of enzyme reactions by this means.

Further Reading

Grimstone, A.V. (1968). *The Electron Microscope in Biology*. Arnold, London.
Haggis, G.H. (1966). *The Electron Microscope in Molecular Biology*. Longmans, London.
Hall, C.E. (1966). *Introduction to Electron Microscopy*, 2nd ed. McGraw–Hill, New York.
Kay, D.H. (1965). *Techniques for Electron Microscopy*, 2nd ed. Blackwell, Oxford.
Meek, G.A. (1971). *Practical Electron Microscopy for Biologists*. Wiley–Interscience, London.
Pease, D.C. (1964). *Histological Techniques for Electron Microscopy*. Academic Press, New York.
Sjostrand, F.S. (1967). 'Instrumentation and Techniques', in *Electron Microscopy of Cells and Tissues*, Vol. 1. Academic Press, New York.
Thornton, P.R. (1968). *Scanning Electron Microscopy*. Chapman and Hall, London.

PROBLEMS

11.1 (a) Calculate the momentum, velocity and wavelength of an electron which has been accelerated through a potential difference of 50 kV.

(b) If the electron then passes through a uniform magnetic field of 0·1 T directed at right angles to its path, what will be the radius of curvature of the resulting circular motion?

11.2 Calculate the focal length of a 'thin' magnetic electron lens for 50 kV electrons, assuming that the magnetic field along the lens axis has a constant value of 0·5 T over a distance of 2 mm, and is zero at points on the axis outside this distance.

11.3 The magnetic objective lens of an electron microscope has a 30 μm diameter objective aperture and a focal length of 5 mm. The spherical aberration constant, C_s, is 0·003 m and the chromatic aberration constant, C_c, is 0·004 m. The accelerating voltage of the microscope fluctuates by up to 2 V in an average of 50 kV.

Estimate the resolution limits imposed separately on the image by spherical aberration, chromatic aberration and diffraction at the objective aperture. Which of these has the worst effect on the resolution, and what could be done about it?

11.4 A tissue section which has been cut for examination in the electron microscope is a square with sides 1 mm long. If the whole section is to be examined at a magnification of × 50 000, what is the corresponding area of the image that will need to be surveyed? (10^4 $m^2 = 1$ hectare $= 2·47$ acres.)

11.5 Prove that the mass per unit area of a given region in a specimen is equal to the density of the region multiplied by its thickness.

A spherical virus of diameter 20 nm and density $1\ 300\ \text{kg m}^{-3}$ lies on a thin carbon supporting film which covers the perforations of a specimen grid in the electron microscope. Estimate the ratio, I_v/I_c, of the image intensities corresponding to the centre of the virus (I_v) and a clear piece of carbon film (I_c) if the effective value of S_t is $2 \times 10^4\ \text{m}^2\ \text{kg}^{-1}$.

CHAPTER 12

Radiation Physics

In this chapter we shall discuss the properties and usefulness of radioactive materials, that is materials in which the atomic nuclei emit particles or electromagnetic radiation. The type of emission from a particular atom does not, in general, depend upon its environment, so radioactive atoms incorporated into biological cells will continue to emit particles or radiation in the same manner as isolated atoms of the same type. By detecting the origin of the radiation in a cell it is possible to follow the movement of a particular atom and hence to discover something about the chemical behaviour of the cell at the molecular level.

Certain types of emission from radioactive materials can cause damage to living cells. It is therefore important both for the interpretation of experimental results and for the health of the experimenter to be aware of the damaging effects and to take the necessary precautions.

12.1 Structure of Nuclei

The modern concept of the atom is a positively charged nucleus surrounded by a number of electrons sufficient to maintain overall charge neutrality. The nucleus consists of a number of particles called *protons* and *neutrons*. Protons have a positive charge equal in magnitude to the charge on an electron while the neutron has no charge. Clearly, to maintain zero overall charge within the atom, the nucleus must contain as many protons as there are electrons orbiting the nucleus. The number of protons, Z, in the nucleus of the atoms of an element is also called the atomic number of the element.

The modern view is that protons and neutrons are different excited states of a single particle, the *nucleon*. Under suitable conditions it is possible to transform a proton into a neutron, and vice versa. The particles within the nucleus are very strongly bound together. The binding forces cannot be electrostatic since only positive charges exist in the nucleus and these would repel each other; electrostatic forces, in any event, could not explain the retention of chargeless particles by the nucleus. A special force, the strong nuclear force, is required to explain the binding together of nucleons. This latter force has a very short effective range, about 10^{-15} m (i.e. about the size of the nucleus), is stronger than the electrostatic forces of repulsion between protons and is independent of charge, so that the same attractive force occurs within the nucleus between two neutrons, two protons or a neutron and a proton.

Since the arrangement of electrons in an atom determines its behaviour in the majority of chemical reactions, its proton number specifies the element of which the atom forms a part (i.e. whether the atom be nitrogen, oxygen or any other element). Two atoms containing the same number of protons in their nuclei, that is two atoms of the same element, need not have the same number of neutrons. Atoms with the same number of protons but which differ in the number of neutrons contained by their nuclei are called *isotopes*. To distinguish between isotopes in chemical formulae superscripts are used to indicate the total number of nucleons in the nucleus. This number is frequently termed the *mass number*, A, of the atom. Thus hydrogen, which has a single proton for its nucleus, would be written ^{1}H to distinguish it from deuterium, which contains a proton and a neutron in its nucleus and is written ^{2}H. Tritium, which has a nucleus consisting of one proton and two neutrons is represented by the symbol ^{3}H. Some authors write the symbol for the element followed by the superscript, thus, H^1, H^2, H^3, but in the present text the superscript will precede the symbol.

The masses of nuclear particles are usually given in atomic mass units (a.m.u.), which are currently defined in terms of the mass of the isotope of carbon with mass number 12. One atomic mass unit is equal to 1/12 of the mass of an atom of ^{12}C. One mole of ^{12}C, that is 6×10^{23} atoms, has a mass of 0·012 kg. One a.m.u. is therefore

$$\frac{1}{12} \times \frac{0·012}{6 \times 10^{23}} = 1·66 \times 10^{-27} \text{ kg.}$$

The proton has a mass of 1·007595 a.m.u., while that of the neutron is 1·008987 a.m.u. It might be expected that the mass of a nucleus would be equal to the sum of the masses of its constituent particles, but this is not found to be so. The nuclear mass is generally less than the sum. To account for this difference in mass use must be made of the equivalence of mass and energy. Thus, if Δm is the decrease in mass when a number of particles combine together to form the nucleus of an atom, an amount of energy $\Delta E = \Delta m\, c^2$, where **c** is the velocity of light, is released during the process. This amount of energy represents the *binding energy* of the particles in the nucleus. To disrupt the nucleus an amount of energy equal to the binding energy must be supplied to it. This energy is sufficiently large that nuclei are, generally speaking, stable structures. Although the figure is somewhat variable for atoms of low mass number the binding energy per nucleon is in the region of 10^{-10} joules for the majority of atoms.

12.2 Radioactivity

Radioactive isotopes emit particles which are often accompanied by electromagnetic radiation of short wavelength. An isotope which is naturally radioactive normally emits either an alpha(α)–particle, which consists of two protons and two neutrons and is therefore the nucleus of helium, or a beta(β)–particle, which is an electron. The radiation which may accompany the particle is called gamma(γ)–radiation.

Most naturally occurring radioactive isotopes have large proton numbers, usually greater than 81 (corresponding to the element thallium), although some (e.g. ^{3}H, ^{14}C, ^{90}Sr) are much lighter than this. It is possible to create artificially a radioactive isotope of every known element, although particles other than α– and β–particles may be emitted by them.

The particles emanating from a radioactive isotope originate in the nucleus of the atom. The removal of a charged particle from the nucleus will produce a change in the proton number of the atom. Thus a radioactive disintegration results in the production of a new atom. In a radioactive process involving the emission of an α–particle from the nucleus of an atom with proton number Z and mass number A a new atom with proton number $Z-2$ and mass number $A-4$ is formed. As an example, radium, which is a solid with $Z = 88$ and $A = 226$, emits an α–particle to form the rare gas radon ($Z = 86$, $A = 222$). Chemical tests on the two types of atom show that a nuclear transformation has taken place. When an atom (proton number Z, mass number A) emits a β–particle the new (daughter) atom has a proton number of $Z+1$ but the same mass number as the parent atom. Thus carbon ($Z = 6$, $A = 14$) disintegrates by emitting a β–particle to form nitrogen ($Z = 7$, $A = 14$). It is perhaps worth noting at this point that to satisfy the laws of conservation of energy and angular momentum another particle, the neutrino (symbol v), accompanies the electron in β–particle emission.

The rate at which radioactive material disintegrates (or decays) is almost independent of all physical and chemical conditions. Consider a large number of atoms of a particular radioactive isotope. The average number, dN, that will decay in a small time dt is proportional to the number, N, of atoms present at time t. That is

$$-dN = \lambda N \, dt, \tag{12.1}$$

where λ is a constant for a particular isotope. The negative sign appears because the number of original atoms is decreasing. By integrating equation (12.1) the following result is obtained:

$$N = N_0 \, e^{-\lambda t} \tag{12.2}$$

where N_0 is the number of original atoms present at $t = 0$. The rate of decay of a radioactive isotope therefore follows an exponential law (Figure 12.1).

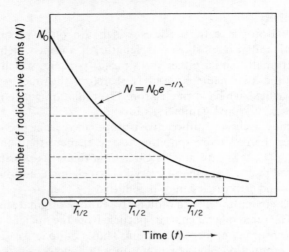

Figure 12.1. Diagram showing the exponential dependence of the rate of decay of a radioactive isotope. The time ($T_{\frac{1}{2}}$) required for half of the original number of atoms to decay is independent of the original number

For practical reasons (e.g. the disposal of radioactive waste) it is often necessary to know how long it will take a particular mass of radioactive material to transform itself into its non–radioactive product. From equation (12.2) and Figure 12.1, it is clear that N only becomes zero at infinite time, so that complete transformation never occurs, but after a certain period of time the level of activity becomes very low. To assist in the computation of decay times a quantity known as the half–life is quoted for radioactive materials. The half–life is the time taken for one–half of the original atoms to decay and can be found from equation (12.2) by putting N equal to $N_0/2$. The half–life $T_{1/2}$ is found to be

$$T_{1/2} = \frac{\ln 2}{\lambda} = \frac{0\cdot 693}{\lambda}.$$ (12.3)

If the half–life is known for an isotope its decay constant, λ, can be computed, thus permitting the evaluation of any required decay times. Notice that as a consequence of the exponential decay law (equation 12.2) the time taken for one–half of a certain mass of radioactive material to become transformed into its product is independent of the mass. This means that it will take as long for one milligramme of radium to degenerate to half a milligramme as it will for one kilogramme of radium to degenerate to half a kilogramme (Figure 12.1). Values for half–lives vary considerably among the radioactive isotopes. Some examples are given in Table 12.1.

TABLE 12.1
Values for half–lives for some radioactive isotopes

Isotope	Half–life
Thorium	$1{\cdot}39 \times 10^{10}$ yr
Carbon 14	5 580 yr
Strontium 90	25 yr
Cobalt 60	5·3 yr
Radon	3·82 days
Sodium 24	15·0 h
Polonium 84	3×10^{-7} s

The activity of a radioactive sample can be expressed in terms of its rate of decay, that is the number of disintegrations per second in the sample. The units in which activity is measured are called curies (Ci), one curie being $3{\cdot}7 \times 10^{10}$ disintegrations per second. This unit is used for all types of nuclear disintegration, irrespective of the particle emitted. An activity of one curie represents a very strong radioactive source, so that it is more usual to encounter milli– and micro–curies in connection with sources commonly used in the laboratory.

12.3 Nuclear Fission

Nuclear fission is of immense practical importance in the modern world since large quantities of energy can be produced and harnessed during the process. When the nucleus of a heavy atom is bombarded with neutrons, it may absorb one of them and thereby become unstable. The unstable nucleus disintegrates into two nuclei of comparable masses with the release of large quantities of energy. In the fission of certain isotopes (e.g. uranium) neutrons are released in addition to the energy. Each of these neutrons is itself capable of initiating a further fission so that a chain reaction can be set up. The fission of uranium 235 may proceed according to the following equation :

$$\ _0^1 n + \ ^{235}U \rightarrow (\ ^{236}U) \rightarrow \ ^{141}Ba + \ ^{92}Kr + 3\ _0^1 n + Q,$$

where $_0^1 n$ represents a neutron. In this equation (^{236}U) represents the un–stable nucleus while Q is the amount of energy released. This energy is equivalent to the difference in mass between the final and the initial particles.

In this particular case the energy released during the fission of a single nucleus is about 3 nanojoules, a figure several million times greater than the energy involved in a combustion process.

The possibility of sustaining a chain reaction depends upon the production of several neutrons by each disintegrating nucleus. A chain reaction may build up, remain steady or die down according to the number of neutrons that remain within the material to continue fission processes; some nuclei escape through the surface of the material while others are absorbed in non–fission processes within the material. It can be shown that there is a *critical size* for fissionable material below which it is impossible to sustain a chain reaction.

A device in which the rate of production of energy by fission processes is controlled, usually by means of inserting material capable of absorbing neutrons, is called a nuclear reactor. The product nuclei of fission processes are often radioactive; in fact, one purpose of a nuclear reactor is to produce radioactive isotopes. To prevent harmful radiation from reaching people outside the reactor suitable shielding, usually in the form of solid concrete, must be provided. When an atomic bomb explodes no shields exist to stop the emission of harmful radiation. The radioactive material which is dispersed in the air following the detonation of an atomic bomb is referred to as the *fallout*. It is to prevent the accumulation of fallout in the atmosphere that the majority of bomb test detonations occur underground.

12.4 Nuclear Fusion

In a nuclear fusion process at least one of the products of the reaction has a mass greater than any of the initial reacting nuclei. Fusion is of scientific importance because it is the source of light and heat energy released by the stars. In fact, energy is released in a fusion process only if the sum of the mass numbers of the reacting nuclei is less than 60. For fusion processes involving a greater value for this sum, energy must be provided in order that the reaction may occur.

Certain widely accepted theories of stellar evolution suggest that all known atomic nuclei are created by fusion processes in the later stages of a star's life. When a star explodes (a so–called supernova explosion) the nuclei become widely distributed in space. It is possible that the complex atomic and molecular organization of living matter here on earth had its origin in the chaotic emanation of nuclei from an ancient exploding star.

Among the lighter nuclei, fusion reactions involving hydrogen and helium nuclei are those that provide the most energy. An example of such a reaction is

$$_1^2H + _1^3H \longrightarrow _2^4He + n.$$

This is one of the main reactions occurring in a thermonuclear weapon.

Because the deuterium and tritium nuclei must approach each other very closely for the reaction to occur, temperatures in the region of 60 million K are necessary to give the particles sufficient energy to overcome their mutual repulsion. Temperatures of this order can be provided by fission reactions. Most of the fallout from a hydrogen bomb originates from the fuel used in the fission reaction needed to produce the temperature required for the more energetic fusion reactions.

12.5 Radiation Detectors

It is relatively easy to detect the presence of charged particles or of gamma rays emanating from a radioactive source, since these types of radiation are capable of ionizing or exciting atoms. In the ionizing process an electron in an atom receives sufficient energy from the impinging radiation to move it outside the sphere of influence of its parent nucleus, thereby creating a positively charged ion. In an excited atom one or more of the bound electrons have orbitals such that the potential energy of the atom is greater than that when the atom is in its ground state. Particles possessing no charge generally produce little or no ionization or excitation, and their presence must be detected by rather indirect methods.

There are many types of detector capable of indicating the presence of charged particles, but in the biological field the instruments most generally used are the Geiger–Müller counter, scintillation counters, solid–state detectors and photographic emulsions.

12.5.1 *The Geiger–Müller Counter*

In the Geiger–Müller counter a high voltage (usually in the range from 400 V to 500 V) is applied between two electrodes placed in a gas, at a pressure somewhat less than atmospheric pressure. An electronic detector records the passage of any current flowing between the electrodes. The

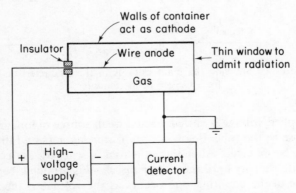

Figure 12.2.   Diagrammatic form of a Geiger–Müller counter

majority of Geiger–Müller counters have cylindrical gas chambers, the wall of which acts as one electrode (the cathode) while a wire anode runs along the axis of the cylinder (Figure 12.2). At one end of the cylinder a thin sheet of material (often aluminium or mica) allows radiated particles to enter the chamber where ionization of the gaseous atoms occurs.

Generally an α– or β–particle will ionize a relatively small number of atoms, thereby producing several positively and negatively charged particles (gaseous ions and electrons). Because of the electric field in the tube of the Geiger–Müller counter, positively charged ions will move towards the cathode while negatively charged electrons will move towards the anode. If the applied voltage is made sufficiently high, these ions and electrons will become sufficiently energetic to ionize further atoms, the ions of which may produce still further ionization. In the Geiger–Müller counter the voltage is such that the products of ionization of a single atom by a particle entering the chamber can give rise to the ionization of about 10^8 further atoms. The charged particles are collected by the anode and cathode and thus produce a current in the external circuit.

The electronic detector in the circuit is a device which will indicate a relatively large pulse of current but which will ignore small pulses. The detector records the number of pulses that flow rather than their magnitude and is called a scaler. Figure 12.3 shows how the counting rate changes

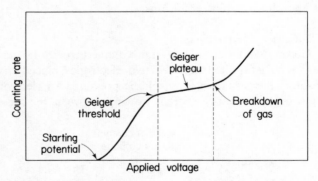

Figure 12.3.   Showing the main features of the characteristic curve for a Geiger–Müller counter

when the applied voltage is varied when a weak source of ionizing radiation is placed close to the tube. Below a certain *starting potential* the pulses are too small to be detected. After this potential is reached, the counting rate increases quite sharply until the *Geiger threshold* is attained, beyond which little increase in the counting rate occurs as the voltage is increased. This relatively flat region of the curve is called the *Geiger plateau*. The plateau terminates at a voltage that is sufficiently high to produce a continuous discharge in the gas in the absence of any radiation.

The counter is always operated at a voltage lying in the middle of the plateau, since in this region small fluctuations in the applied voltage have only a small effect on the counting rate. Typically a counter with an operating voltage of 800 V has a plateau about 300 V in extent.

There is a natural limit to the counting rate set by the time taken for the discharge due to a given incident particle to be completed. Clearly, if a particle enters the chamber before the discharge produced by the preceding one is over, it will not be registered. The discharge time is shortened by the inclusion of small quantities of a polyatomic organic gas (e.g. methane) or a halogen in addition to the gas to be ionized (this is often argon). An alternative method for reducing the discharge time is to lower the applied potential to a value below the starting potential immediately after a particle has entered the tube and, when the discharge is over, to raise it in preparation for the next particle. Fairly sophisticated electronic circuitry is needed if the latter method is used. The processes described in this paragraph are known as *quenching*.

If particles were to reach the counting tube at a uniform rate, a maximum of 5 000 pulses per second could be recorded with a properly quenched tube. The emission from radioactive sources is, in fact, a random process, so that the maximum practical counting rate is less than that predicted and is often less than 1 000 pulses per second. Corrections are necessary to allow for the possibility that two successive particles may arrive at the tube with a time interval sufficiently short that only one pulse is produced. The correction necessary can be calculated statistically and becomes smaller as the quenching time is reduced or the counting rate made slower.

Geiger–Müller counters cannot differentiate between α– and β–particles or γ–rays. They are generally used for the detection of β–particles and γ–rays because of the technical difficulty of making a sufficiently thin window for the passage of the less penetrating α–particles. These counters are generally used for comparison work, in which case it is not necessary to know the *geometric efficiency* of the system. The geometric efficiency is simply that fraction of the total number of particles emitted from the source which actually enter the tube. In comparative work the geometry must remain the same for all sources. To make absolute measurements of the activity of a source a knowledge of the geometry is required. In this case a standard radioactive source would be used to calibrate the instrument.

12.5.2 Solid–state Detectors

Solid–state detectors are constructed from semiconducting materials like silicon and germanium (Chapter 6) and are similar in their action to the gaseous ionization chamber, of which the Geiger–Müller counter is an example. A semiconductor *p–n* junction detector is shown in Figure 12.4. An energetic radioactive particle which enters the detector produces ionization, so that electrons are moved from the valence band to the conduction

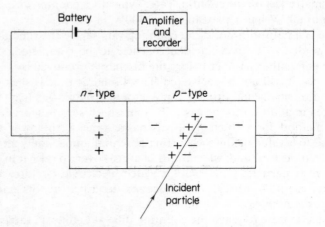

Figure 12.4. The main features of a semiconductor *p–n* junction detector

band of the material. For each electron which moves to the conduction band, a positively charged hole remains in the valence band. As explained in Chapter 6, if a reverse bias is applied across the junction, the electrons and holes produced in this way will move through the material thus generating a current which may be amplified and recorded. Thus, as in the Geiger counter, the passage of a radioactive particle is registered as a current pulse. The characteristics of the device are such that the counting rate, which may exceed 10^6 pulses per second, is limited by the electronics used to amplify and record the pulse rather than by the semiconductor. The response of the instrument depends upon the energy of the incoming particle and may be used, with the appropriate circuits, to count separately particles of different energies in an inhomogeneous beam.

A solid–state counter has a small physical size and can be made to be very robust. Because of this it is becoming widely used for medical and biological measurements where previously miniature Geiger–Müller counters or scintillation counters were used. A further advantage over the Geiger–Müller counter is that a solid–state detector operates at low voltages, usually in the region of 50 V but sometimes as low as 10 V. This consideration is important for safety, in such applications as the localization of brain tumours as described in Section 12.6.

12.5.3 *Scintillation Counters*

In a scintillation counter use is made of the fact that certain materials, called phosphors, emit a small pulse of light when ionizing radiation passes through them. This is because atoms of the materials can be excited by the radiation; when the atoms return to their ground states, visible or ultraviolet light is emitted (Chapter 7). The light can be detected by a photo–multiplier tube (Chapter 6), by means of which a light signal is transformed into an electrical pulse which can be amplified and used to obtain a measure of the light intensity. A schematic diagram of a scintillation counter is shown in Figure 12.5.

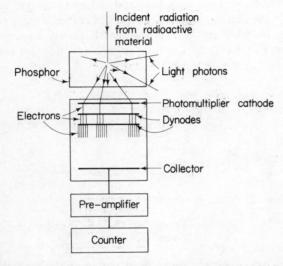

Figure 12.5.   Diagrammatic form of a scintillation counter

The type of phosphor used in a scintillation counter depends upon the type of radiation to be detected. Zinc sulphide, for example, is a good phosphor in that it converts about one–quarter of the energy of the incident radiation into light energy. It is, however, not very transparent, so that only thin layers of this material can be used if all the light pulses produced in the phosphor are to reach the photomultiplier. Such a phosphor is suitable for radiation which will be stopped by very thin pieces of material, and is therefore used to detect α–particles. Sodium iodide 'activated' by thallium is widely used as a phosphor for detecting β– and γ–radiation, although other materials are used for special purposes.

The number of light photons produced in the phosphor is determined by the amount of energy transferred from the incident radiation to the phosphor, which thus governs the size of the pulse produced by the pre-amplifier. The energy absorbed by the crystal is related to the energy of the incident radiation, so that a calibrated scintillation counter can be used to measure this energy. Suitable counter circuits have been devised that allow high–energy particles to be monitored in the presence of low-energy particles, and vice versa; thus, it is sometimes possible to measure the activity of one isotope in the presence of another if the two emit radiation of differing energies.

In general, the phosphor is maintained in close optical contact with the cathode of the photomultiplier but in certain circumstances (e.g. where it is desirable to keep the dimensions of the detector itself small) the phosphor and photocathode may be separated and linked by a light guide.

To measure the activity of small samples of material such as are obtained when samples of blood are used for medical diagnosis, a special well–shaped crystal is used (Figure 12.6). Liquid phosphors, which are usually complex

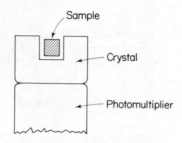

Figure 12.6.   Well–shaped crystal used for certain scintillation counter measurements

phenol derivatives, are sometimes used to count the particles from low–energy β–emitting sources such as ^{14}C. The radioactive material is mixed with the phosphor and placed in a suitable container close to the photocathode.

The scintillation counter has advantages over the Geiger–Müller counter in that its phosphor detector is more rugged than the Geiger–Müller tube and that it is capable of a higher counting rate for β– and γ–radiation.

12.5.4 *Photographic Emulsions for Particle Detection*

A photographic emulsion consists of silver halide grains embedded in a film of gelatin. The grains can be activated by ionizing radiations from radioactive materials and can then be converted to silver by a suitable chemical known as a developer. Special emulsions have been developed for recording the passage of ionizing particles; since the particles generally come from radioactive nuclei, the emulsions are called *nuclear emulsions*.

The grains in a nuclear emulsion are usually silver bromide and are more densely packed than in an emulsion intended for photography using light. The high density ensures that as many radioactive particles as possible will be stopped by the emulsion.

Because nuclear emulsions are designed to detect the passage of very small particles, the grain size in them tends to be smaller than in other photographic emulsions. The smaller the grain size, the shorter will be the time of transit of a particle in the crystal and hence the smaller will be the energy liberated to activate the crystal. This situation is often referred to by saying that a nuclear emulsion with small grain size has a low sensitivity. It is important that the grain size and sensitivity of a given nuclear emulsion be uniform since it will be used to obtain quantitative results concerning the emission of particles from radioactive materials.

12.5.5 *Background*

In any quantitative measurement of radioactivity it is always necessary to take account of the *background*. When any system of detection is used, responses are always possible from sources other than the radiation in which we are interested; signals from these sources give rise to the back–ground level of recorded radioactivity. Care must therefore be taken to make control experiments to determine the amount of background, so that it can be subtracted from the signal obtained from the experimental material.

12.6 The Use of Radioactive Isotopes in Biology

Radioactive isotopes can be used in biological experiments because living cells do not discriminate between isotopes of a particular element. Since the electronic structure, and therefore the chemistry, of one isotope is the same as another, a cell will make use of carbonaceous material containing radioactive ^{14}C equally as well as similar material containing only non-radioactive isotopes of carbon. Because the presence of a radioactive subs-tance can be detected with suitable instruments, it is possible to 'label' compounds used during cellular metabolism by incorporating radioactive isotopes into them. It is thus possible, at least in principle, to trace the passage of such a compound as it becomes involved in the life of a cell. This type of compound is often called a *tracer*, for obvious reasons.

Studies with radioactive isotopes are generally of two types: (a) those in which overall uptake of material by tissues or single cells is measured and (b) those where it is required to locate that part of a cell containing the radioactive material. Measurements falling into category (a) are usually made with a Geiger–Müller counter or with a scintillation counter, while photographic emulsions are the detectors used for studies of type (b).

Uptake of material by animal tissue can be investigated *in vivo* or by *post mortem* examination according to the type of experiment. Radioactive iodine, for instance, has been widely used in both the *in vivo* study of the behaviour of the human thyroid gland and in the treatment of disorders in this type of gland. In a specific examination of this type, the subject is given orally about 25 μCi of radioactive iodine (in the form of a solution of sodium iodide). The action of the thyroid gland concentrates the iodine, which emits γ–rays sufficiently energetic to penetrate the tissue between the gland and an external detector. By measuring the signal, and hence the amount of iodine concentrated by the thyroid, defects of the thyroid gland can be diagnozed.

In vivo localization of brain tumours is achieved by giving the patient an injection of 1 mCi of ^{32}P in the form of sodium phosphate. The degeneration of the blood/brain barrier around tumour tissue combined with the rapid generation rate of tumour cells means that there will be a large difference between the ^{32}P uptake of tumour cells and that of the normal brain tissue. A small probe inserted into the region of the tumour can thus be used to locate it accurately.

The general technique employed for studying the behaviour of single cells involves suspending the cells in a medium containing the radioactive material. After the appropriate period of time the cells are centrifuged and thoroughly washed with non–radioactive solutions and their activity measured. As an example, the uptake of phosphorus by cells has been studied by using sodium dihydrogen phosphate as the source of phosphorus. Some of the phosphate contained the radioactive isotope ^{32}P, so that its uptake by the cells could be measured and correlated with other factors, such as available chemical energy.

Investigations of type (b) are normally classed under the heading *auto–radiography*, although the term radioautography is also used. To find the position of radioactive material within the tissue or a cell it is necessary to place a photographic emulsion very close to the cell itself and to allow the

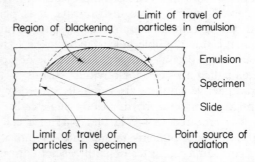

Figure 12.7. Showing the arrangement necessary for autoradiography, and the region of the emulsion affected by a point source of radiation

particles emitted by the isotope to activate the grains in the emulsion. To achieve this, a microscope slide bearing the biological material is coated with the photographic emulsion (Figure 12.7). After sufficient time has elapsed (the time is usually determined by trial and error for a particular set of conditions), the emulsion is processed, with the cellular material still in position, and the slide is examined with a microscope. Dark regions will be observed above any part of the cell containing the radioactive isotope. The ability to separate two closely spaced radioactive sources in the cell (i.e. the resolution) is limited by the finite thickness of the cell section and photographic emulsion and also by the finite range of the emitted particles. Thus, a point source of radiation will give rise to a circular area of blackening having a finite radius (Figure 12.7).

The electron microscope (Chapter 11) has also been used with auto–radiography to locate radioactivity in those parts of cells too small to be seen in the light microscope. The full potential of the combination has not yet been realized, however, since the maximum resolution obtainable with nuclear emulsions is about 700 Å whereas the electron microscope is capable of resolving objects separated by distances of 5 Å to 10 Å. This situation will clearly improve as nuclear emulsions are refined.

In the autoradiographic method quantitative estimates of the activity of the material producing the blackened emulsion are obtained by counting the number of blackened grains in the emulsion. The interpretation of the results is complex and requires a consideration of many factors, such as the length of exposure and the proximity of neighbouring labelled material in the specimen. The background must clearly be subtracted from an experimental count to obtain the effect due to the source of interest. This may be achieved by using a control slide of similar material, but containing no radioactive isotope, or by observing the emulsion at a region on the experimental slide which is remote from the radioactive tissue.

Autoradiography in combination with light microscopy has been used to show that during the growth of flagella in protozoa, material is added to the tip of the flagellum rather than to its base, the site at which we might expect addition of new molecules. Electron–microscope autoradiography has shown that in mitochondria there is a continual turnover of protein, so that these cell components are not replaced as complete entities as had previously been speculated. The half–life of the mitochondrial protein is found to be about nine days.

12.7 Radiation Damage

Since the particles emitted by radioactive sources are capable of producing ionization, the molecules within a cell can be ionized if the cell is exposed to such radiation. This changes the charge distribution on the molecule which may now react chemically with other cell components to form new molecules which may be harmful to the cell. If enough of the essential molecules within the cell are modified in this way, the cell itself will die. When humans are exposed to ionizing radiations, which include X–rays as well as the products of radioactivity, certain of their cells may die, producing radiation–sickness or even burns if the radiation is sufficiently intense. Often the effects of radiation (e.g. skin cancer or leukaemia) are not noticed for several years. In addition to direct action upon the cells of an individual, radiation can produce genetic effects, which appear in children of the irradiated person and may not manifest themselves for several generations.

When considering the effects of radiation on biological specimens it is more convenient to use units which express the effects of radiation on matter than to use the curie, which is a number of particles emitted by a source in a given time. It is often useful to be able to express quantitatively the amount of radiation to which a particular region has been subjected without specifying the material within the region. In this case the ionizing effect of the radiation dose on a standard material, namely air, is used and is called the *exposure*. The unit of exposure is the *röntgen* (**R**), which would produce by ionization $2 \cdot 58 \times 10^{-4}$ coulombs of charge (measured for the positively *or* the negatively charged ions) per kilogramme of air. For a particular radiation the exposure–rate (measured in $R\ s^{-1}$) is proportional to the intensity of the beam.

The quantity of radiation with which a particular region has been treated can also be expressed in terms of the energy absorbed from the radiation per unit mass of material. This is called the *absorbed dose* and is measured

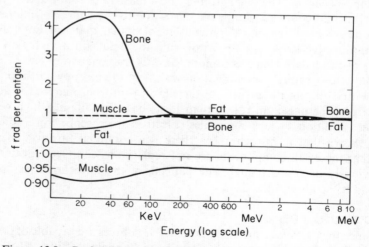

Figure 12.8. Ratio, *f*, between absorbed dose and exposure for some biological materials. (Reproduced with kind permission from M.E.J. Young, *Radiological Physics*, H.K. Lewis, London, 1967)

in *rads*, where one rad is equivalent to the absorption of ten millijoules of energy by a kilogramme of material. An impression of the size of this unit can be obtained by considering that about four kilojoules of energy are required to raise the temperature of one kilogramme of water by one kelvin.

The absorbed dose is related to the exposure, but for a given exposure the absorbed dose will depend on the material irradiated and also upon the energy of the radiation. For example, bone will absorb more rads per röntgen than muscle or fat from a beam of energy 20 keV, but for beam energies greater than about 200 keV the absorption per röntgen of the three types of material is approximately the same and remains constant at about one rad per röntgen as the beam energy increases (Figure 12.8).

It is also convenient to have yet another measure of the effect of radiation on matter, since to produce the same effect in tissue requires a different dose of one type of radiation than of another. For example, ten rad from X–rays produce the same damage as one rad from α–particles. It has become customary to introduce the term *dose equivalent*, such that equal dose equivalents of different radiation produce the same effect. A dose equivalent, which is measured in *rems*, for any ionizing radiation is the absorbed dose multiplied by a factor which depends on the radiation and sometimes on the irradiated tissue. Thus for X–rays the factor is unity while for α–particles, it is ten for general tissues but thirty for the lens of the eye.

The ionization of atoms by radiation within a living cell can have profound effects on the chemistry, and consequently the future activity, of the cell. Experiments have been carried out, *in vitro*, on a variety of molecules which occur in living matter to find out the effects of radiation. Proteins, for example, may either break down to form smaller molecules or produce larger molecules by means of cross–linkages between chains. Enzymes in aqueous solution are readily inactivated by radiation but are more resistant to damage in the presence of their substrates. When deoxyribosenucleic acid is irradiated in aqueous solution breakage of the chains occurs. It is important to remember that the effects *in vitro* may well be different from those *in vivo*, where the chemical environment may not be the same.

It is not usually possible to relate the large–scale effects of radiation (death of cells or organisms, etc.) to a particular chemical change, although techniques have been devised by means of which it is possible to estimate the size of a region which must have been damaged to produce the observed gross effect. Perhaps the most important effect of radiation on cells is the prevention or impairment of reproduction. Much evidence exists to support the idea that this effect is a result of chromosome damage.

Since ionizing radiation is capable of killing cells, much effort has been directed towards killing harmful cells (e.g. tumour cells) in the human body. This is the technique known as *radiotherapy*. The main problem associated with this technique is that normal cells are often as sensitive to the radiation as the abnormal ones.

Because radiation is harmful to human cells it is essential that anyone working continuously near sources of radiation (e.g. an electron microscope, which can produce X-rays if not shielded properly) should carry a device for measuring the exposure to which they have been subjected. The usual device is a film-badge, which consists of a small piece of photographic film partially covered by a thin sheet of metal (usually lead). The whole assembly is covered so that light cannot reach the film and is mounted in a frame which can be fastened to clothing. When the film is developed the degree of blackening is a measure of the radiation exposure. Penetrating radiations produce blackening under the metal shield, while the less penetrating radiations produce blackening only in the unshielded region. The International Commission on Radiological Protection has laid down maximum doses both for members of the public and for persons whose work involves the use of radioactive isotopes. Examples of the maximum dose limits are given in Table 12.2.

TABLE 12.2
Dose limits for individuals

| Region of exposure | Occupational exposure | | Members of the public |
	In 13 weeks (rem)	In 1 year (rem)	In 1 year (rem)
Gonads and red bone marrow	3	5	0·5
Abdomen of women of reproductive capacity*	1·3	5	
Skin, thyroid, bone	15	30	1·5 in thyroid of children under 16 years of age
Other single organs	8	15	1·5
Extremities	38	75	7·5

* Note that following confirmation of pregnancy the dose to the foetus during the remainder of the pregnancy should be restricted to 1 rem.

For occupational exposure of individuals below 18 years of age, the dose in any one year must not exceed 5 rems and the dose accumulated up to the age of 30 must not exceed 60 rems.

In view of the harmful effects of radiation described earlier, and since the interaction of radiation with matter is poorly understood, it is essential that extreme caution be exercised by persons working habitually with instruments producing radiation or with radioactive isotopes. Adequate shielding must be used so that the dose absorbed by an individual is kept as low as possible.

Further Reading

Boursnell, J.C. (1958). *Safety Techniques for Radioactive Tracers*. Cambridge University Press, London.

Bradshaw, L.J. (1966). *Introduction to Molecular Biological Techniques*. Prentice–Hall, New Jersey.

Casarett, A.P. (1968). *Radiation Biology*. Prentice–Hall, New Jersey.

Comar, C.L. (1955). *Radioisotopes in Biology and Agriculture*. McGraw–Hill, New York.

Hine, G.J. (Ed.) (1967). *Instrumentation in Nuclear Medicine*. Academic Press, New York.

Kamen, M.D. (1951). *Radioactive Tracers in Biology*. Academic Press, New York.

Lambie, D.A. (1964). *Techniques for the Use of Radioisotopes in Analysis*. Spon, London.

Lea, D.E. (1955). *Actions of Radiations on Living Cells*, 2nd ed. Cambridge University Press, London.

Morgan, J. (1969). *Introduction to University Physics*, Vol. 2, 2nd ed. Allyn and Bacon, Boston.

Putman, J.L. (1960). *Isotopes*. Penguin, London.

Rogers, A.W. (1967). *Techniques of Autoradiography*. Elsevier, Amsterdam.

Sharpe, J. (1964). *Nuclear Radiation Detectors*, 2nd ed. Methuen, London.

Young, M.E.J. (1967). *Radiological Physics*. H.K. Lewis, London.

PROBLEMS

12.1 (a) The following radioactive isotopes, which are useful in biological tracer work, decay by emitting a β–particle and a neutrino:

$$^{24}_{11}\text{Na}, \quad ^{45}_{20}\text{Ca}, \quad ^{60}_{27}\text{Co}, \quad ^{89}_{38}\text{Sr}, \quad ^{131}_{53}\text{I}.$$

What atom is formed as a result of each decay process?

(b) Investigations of the effect of irradiating biological systems with α–particles are usually carried out using one of the following isotopes, each of which emits only α–particles, as a particle source:

$$^{222}_{86}\text{Rn}, \quad ^{210}_{84}\text{Po}, \quad ^{239}_{94}\text{Pu}.$$

What atom is formed as a result of each decay process?

12.2 The radioactive isotope iodine 131, which is sometimes used to assist a medical diagnosis, has a half–life of 8·04 days. A sample of the isotope, due to be administered to a patient, is known to have an activity of 25 μCi at noon on a certain day. What is its activity at noon twelve days later? How much longer will it be before the activity has dropped to 1 per cent. of the initial value?

12.3 Amino acids, substances essential to the creation and maintenance of life, have been found in the oldest known rocks on earth. The age of these rocks can sometimes be estimated from an examination of the amount of naturally occurring radioactive rubidium 87 which they contain.

(a) In a certain rock sample, 0·700 per cent. of the weight is found to be ^{87}Rb, while 0·035 per cent. is found to be ^{87}Sr, the product of the decay process. If no ^{87}Sr was present when the rock was formed, how old is the rock sample?

(b) In rocks containing fossils of the earliest known complex animals the ratio of ^{87}Sr to ^{87}Rb is 0·009. Estimate the age of the fossils.

(c) What emissions occur when a ^{87}Rb nucleus decays?

(The masses of the nuclei of ^{87}Rb and ^{87}Sr are approximately equal. The half–life of ^{87}Rb is $4·7 \times 10^{10}$ years.)

Solutions of Problems

Chapter 1

1.1 0·19 N at 52·5° to horizontal
1.2 4·90 N; 169.
1.3 (a) 272 J kg^{-1}, (b) 5·4 kW kg^{-1}
For locust: (a) 116 J kg^{-1}, (b) 4·6 kW kg^{-1}
1.4 0·018 s; $K_{critical} = 16\cdot5$ mN s m^{-1}
1.5 90·7 per cent.
1.6 $3\cdot3 \times 10^3$ N m; 336 kg wt

Chapter 2

2.1 $a = 32\ \mu$V K^{-1}, $b = 0\cdot04\ \mu$V K^{-2}
Near $t = 0$, sensitivity dE/d$t = a = 32\ \mu$V K^{-1}
2.2 (b) 1·56 kW, (c) 1·47 kW. Hence (b) gives most power.
2.3 Ratio is $16^{5/4} = 32$.
2.4 (a) 238 W, (b) 3·5 kW
2.5 (a) $1\cdot15 \times 10^{10}$ W, (b) $\lambda_p = 2\cdot9\ \mu$m, (c) 51 W m^{-2}
2.6 (a) 7 W, (b) 378 K or 105°C
2.7 $\Delta T = 56$ K

Chapter 3

3.1 123×10^3 N m^{-2}; greater pressure on concave side
3.2 (a) 84 N m^{-2}, (b) 280 mW
3.3 (a) 62·5 mol m^{-3}; 12×10^{-18} mol
3.4 0·1 m^2 s^{-1}; e.g. glycerin at 20°C to 25°C

Chapter 4

4.1 Due to PO$_4$$^{3-}$ ion, $F = 6\cdot9 \times 10^{-12}$ N
Due to field, $F = 1\cdot6 \times 10^{-15}$ N
4.2 10^{-26} N m

4.3 (a) 177 pF,   (b) 1062 pF
4.4 (a) 2500 A m^{-1},   (b) $3 \cdot 14 \times 10^{-3}$ T,   (c) $3 \cdot 14 \times 10^7$ Wb
4.5 $R = 28 \cdot 8 \, \Omega$;   diameter $= 0 \cdot 094$ mm
4.6 3000 Ω gives most power;   ratio is 1 : 0·88.
4.7 $S = 139$ kΩ

Chapter 5

5.1 (a) 0·1 Wb,   (b) 31·4 V
5.2 300 V
5.3 $\tau = 0 \cdot 25$ ms.   (a) Faithfully,   (b) with distortion,   (c) hardly at all
5.4 500 W
5.5 (a) 6284 Ω,   (b) 0·0253 μF,   (c) 0·1 A
5.6 $X_L = 1600 \, \pi$; $X_C = 2500 \, \pi$;   2830 Ω.
5.7 40 : 1.

Chapter 6

6.1 $5 \cdot 9 \times 10^4$; $2 \cdot 76 \times 10^5$.
6.2 (a) intrinsic,   (b) extrinsic,   (c) acceptors,   (d) p–type,
      (e) electrons
6.3 Negative terminal;   minority ones;   holes.
6.4 $2 \times 10^4 = 20$ kΩ;   2 μA.
6.5 Positive, applying reverse bias to the n–type gate material and
      increasing the extent of the depletion region
6.6 Use, for example, the pnp transistor in a common–collector arrange-
      ment (to give high input impedance) followed by the npn transistor
      in a common–emitter circuit for amplification.

Chapter 7

7.1 Work function $= 3 \cdot 2 \times 10^{-19}$ J; $v = 4 \cdot 83 \times 10^{14}$ Hz; $\lambda = 621$ nm.
7.2 Near infrared associated with molecular vibrations.
      (a) $\cdot 10^{14}$ Hz,   (b) 3333 cm^{-1},   (c) 0·415 eV
7.3 $2 \cdot 7 \times 10^{-18}$ J;   388 kcal mol^{-1}
7.4 251 photons per second; $4 \cdot 02 \times 10^{-11}$ A
7.5 (a) $\tau = 0 \cdot 38$,   (b) O.D. $= 0 \cdot 420$.   Dilution is 1/2.
7.6 0·18 mol m^{-3}

Chapter 8

8.1 (a) 18·8 cm,   (b) 2·26 m
8.2 2·1 m;   0·05

8.3 22 mm
8.4 (a) 5 cm,   (b) × 5

Chapter 9

9.1 $\varepsilon = \pi/4$; $E = 0{\cdot}707\, E_0$.
9.2 0·0273
9.3 $\lambda = 589$ nm;   4 orders;   'slits' have a width which is 1/3 of the distance between them.
9.4 61 μm
9.5 N.A. = 1·30; $D_{min} = 420$ nm
9.6 12·4 μm;   elliptically polarized; $I_{max}/I_{min} = 3{\cdot}00$
9.7 $2{\cdot}18 \times 10^{-8}$;   0·105 rad

Chapter 10

10.1 140 nm
10.2 (a) 0·109,   (b) 0·68
10.3 × 175;   × 14 to give four times the minimum magnification
10.4 0·77;   1·15;   N.A. of condenser should equal that of objective.

Chapter 11

11.1 (a) momentum = $12{\cdot}1 \times 10^{-23}$ kg m s^{-1};
velocity = $1{\cdot}33 \times 10^{8}$ m s^{-1}; $\lambda = 0{\cdot}55 \times 10^{-11}$ m or 0·055 Å,
(b) 0·756 cm
11.2 4·56 mm
11.3 Spherical aberration: 0·81 Å
Chromatic aberration: 4·8 Å
Diffraction   : 9·1 Å (worst)
Increase size of objective aperture to make diffraction limit equal to that then imposed by other aberrations.
11.4 0·25 hectares or 0·62 acres
11.5 Volume of region with area A, thickness t is At and mass is therefore ρAt if density is ρ;   mass/area = $\rho At/A = \rho t$; $I_v/I_c = 0{\cdot}59$

Chapter 12

12.1 (a) $^{24}_{12}$Mg; $^{45}_{21}$Sc; $^{60}_{28}$Ni; $^{89}_{39}$Y; $^{131}_{54}$Xe,
(b) $^{218}_{84}$Pc; $^{206}_{82}$Pb; $^{235}_{92}$U
12.2 8·89 μCi;   41·4 days
12.3 (a) $3{\cdot}31 \times 10^{9}$ years,   (b) $6{\cdot}1 \times 10^{8}$ years,
(c) ^{87}Rb$\rightarrow$^{87}Sr$+\beta+\bar{v}$

Appendixes

Names and symbols of some S.I. units

Quantity	Unit	Symbol
length	metre	m
mass	kilogram	kg
time	second	s
electric current	ampere	A
thermodynamic temperature	kelvin	K
luminous intensity	candela	cd
amount of substance	mole	mol
force	newton	N
energy	joule	J
power	watt	W
frequency	hertz	Hz
electric charge	coulomb	C
electric potential	volt	V
electric resistance	ohm	Ω
electric capacitance	farad	F
magnetic flux	weber	Wb
magnetic flux density	tesla	T
inductance	henry	H
luminous flux	lumen	lm
illumination	lux	lx

Prefixes for S.I. units

The following prefixes are used to indicate decimal multiples of named units. For example 1 MW (one megawatt) is 10^6 watts.

Fraction	Prefix	Symbol	Multiple	Prefix	Symbol
10^{-1}	deci	d	10	deca	da
10^{-2}	centi	c	10^2	hecto	h
10^{-3}	milli	m	10^3	kilo	k
10^{-6}	micro	μ	10^6	mega	M
10^{-9}	nano	n	10^9	giga	G
10^{-12}	pico	p	10^{12}	tera	T
10^{-15}	femto	f			
10^{-18}	atto	a			

Conversion factors

Unit	Symbol	Equivalent
ångstrom	Å	10^{-10} m or 0·1 nm
micron	μm	10^{-6} m
litre	l	10^{-3} m^3 or 1 dm^3
dyne	dyn	10^{-5} N
erg	erg	10^{-7} J
poise	P	$0·1$ N s m^{-2}
gauss	G	10^{-4} T
inch	in	$2·54 \times 10^{-2}$ m
pound	lb	0·4536 kg
atmosphere	atm	101325 N m^{-2}
torr	Torr	133·3 N m^{-2}
millimetre of mercury (pressure)	mmHg	133·3 N m^{-2}
calorie	cal	4·186 J
rad (radiation dose)	rad	10^{-2} J kg^{-1}
rontgen	R	$2·58 \times 10^{-4}$ C kg^{-1}
electron volt	eV	$1·602 \times 10^{-19}$ J

Values of physical constants

Constant	Symbol	Value
acceleration due to gravity*	g	ca. 9·81 m s^{-2}
Avogadro constant	N_A	$6·022 \times 10^{23}$ mol^{-1}
Boltzmann constant	$k = R/N_A$	$1·381 \times 10^{-23}$ J K^{-1}
curie	Ci	$37·0 \times 10^9$ s^{-1}
electronic charge	e	$1·602 \times 10^{-19}$ C
gas constant	R	8·314 J K^{-1} mol^{-1}
'ice–point' temperature	T_{ice}	273·150 K
permeability of a vacuum	μ_0	$4\pi \times 10^{-7}$ H m^{-1} (exactly)
permittivity of a vacuum	ε_0	$8·854 \times 10^{-12}$ F m^{-1}
Planck constant	h	$6·626 \times 10^{-34}$ J s
rest mass of electron	m_e	$9·110 \times 10^{-31}$ kg
rest mass of proton	m_p	$1·673 \times 10^{-27}$ kg
speed of light in a vacuum	c	$2·998 \times 10^8$ m s^{-1}
standard pressure (1 atm)	P	101325 N m^{-2}
Stefan–Boltzmann constant	σ	$5·670 \times 10^{-8}$ W m^{-2} K^{-4}
triple point of water	T_3	273·16 K (exactly)
unified atomic mass constant[†]	m_u	$1·661 \times 10^{-27}$ kg

* Not strictly a constant, since local variations occur over the earth's surface.

[+] One–twelfth of the rest mass of the atom ^{12}C; the relative atomic mass ('atomic weight') of an element is the average rest mass of the atom divided by m_u.

Differentiation and integration

$\int y\,dx$	y	$\dfrac{dy}{dx}$
$\dfrac{x^{n+1}}{n+1}\ (n \neq -1)$	x^n	$n\,x^{n-1}$
$\log_e x$	$\dfrac{1}{x}$	$-\dfrac{1}{x^2}$
$\dfrac{e^{ax}}{a}$	e^{ax}	$a\,e^{ax}$
$\dfrac{a^x}{\log_e a}$	a^x	$a^x \log_e a$
$x \log_e x - x$	$\log_e x$	$\dfrac{1}{x}$
$-\cos x$	$\sin x$	$\cos x$
$\sin x$	$\cos x$	$-\sin x$
$-\log_e(\cos x)$	$\tan x$	$\sec^2 x$
	uv	$u\dfrac{dv}{dx} + v\dfrac{du}{dx}$

Note : differentiation of an expression in one column gives that in the next column to its right, while integration results in a similar movement to the left.

Trigonometric ratios

In the right–angled triangle shown, in which the angle BAC has a value α, the following ratios can be defined:

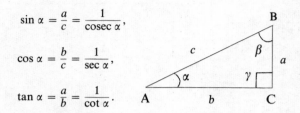

$$\sin \alpha = \frac{a}{c} = \frac{1}{\operatorname{cosec} \alpha},$$

$$\cos \alpha = \frac{b}{c} = \frac{1}{\sec \alpha},$$

$$\tan \alpha = \frac{a}{b} = \frac{1}{\cot \alpha}.$$

For *any* triangle, not necessarily a right–angled one, it holds that

$$a^2 = b^2 + c^2 - 2bc \cos \alpha$$

and

$$\frac{a}{\sin \alpha} = \frac{b}{\sin \beta} = \frac{c}{\sin \gamma}.$$

Other useful relations are:

$$\sin(\theta + \phi) = \sin \theta \cos \phi + \cos \theta \sin \phi,$$
$$\cos(\theta + \phi) = \cos \theta \cos \phi - \sin \theta \sin \phi.$$

If θ is an angle measured in radians,

$$\sin \theta = \theta - \frac{\theta^3}{3!} + \frac{\theta^5}{5!} + \cdots$$

and

$$\cos \theta = 1 - \frac{\theta^2}{2!} + \frac{\theta^4}{4!} + \cdots$$

Index

349